D0240103

nown v.

TSWCH

00016877

Living with Haemophilia

Living with Haemophilia

Fourth Edition

PETER JONES, MD, FRCP, DCH

Director, Northern Regional Haemophilia Service,
Newcastle Haemophilia Centre, The Royal Victoria Infirmary,
Newcastle upon Tyne, United Kingdom;
Consultant Paediatrician and Clinical Lecturer in Child Health,
University of Newcastle upon Tyne, United Kingdom;
Chairman of the World Federation of
Hemophilia Comprehensive Care Committee

LIBRARY
TOR & SW COLLEGE
OF HEALTH / N FRIARY
GREENBANK TERRACE
PLYMOUTH PL4 8QQ

Oxford   New York   Tokyo
OXFORD UNIVERSITY PRESS
1995

Oxford University Press, Walton Street, Oxford OX2 6DP

Oxford   New York
Athens   Auckland   Bangkok   Bombay
Calcutta   Cape Town   Dar es Salaam   Delhi
Florence   Hong Kong   Istanbul   Karachi
Kuala Lumpur   Madras   Madrid   Melbourne
Mexico City   Nairobi   Paris   Singapore
Taipei   Tokyo   Toronto
and associated companies in
Berlin   Ibadan

Oxford is a trade mark of Oxford University Press

Published in the United States
by Oxford University Press Inc., New York

© Peter Jones, 1995

All rights reserved. No part of this publication may be
reproduced, stored in a retrieval system, or transmitted, in any
form or by any means, without the prior permission in writing of Oxford
University Press. Within the UK, exceptions are allowed in respect of any
fair dealing for the purpose of research or private study, or criticism or
review, as permitted under the Copyright, Designs and Patents Act, 1988, or
in the case of reprographic reproduction in accordance with the terms of
licences issued by the Copyright Licensing Agency. Enquiries concerning
reproduction outside those terms and in other countries should be sent to
the Rights Department, Oxford University Press, at the address above.

This book is sold subject to the condition that it shall not,
by way of trade or otherwise, be lent, re-sold, hired out, or otherwise
circulated without the publisher's prior consent in any form of binding
or cover other than that in which it is published and without a similar
condition including this condition being imposed
on the subsequent purchaser.

A catalogue record for this book is available from the British Library

Library of Congress Cataloging in Publication Data
Jones, Peter, 1937–
Living with haemophilia / Peter Jones.—4th ed.
Includes index.
1. Hemophilia.   I. Title.
RC642.J66   1995   616.1'572—dc20   94–33382
ISBN 0 19 263030 X

Typeset by The Electronic Book Factory Ltd, Fife
Printed in Great Britain by
Biddles Ltd, Guildford and King's Lynn

Preface

On September 14th 1990 the first human being in history received gene therapy. She was a little girl living in the United States, and she suffered from a rare inherited abnormality which made her prone to overwhelming infections. She is now thriving and growing and developing normally, and is able to go to school with her friends.

Since 1990 other children have received gene therapy, and the stage is set for a revolution in medicine at least comparable to the introduction of antibiotics in the 1940s. It is an exciting time, and haemophilia is part of the excitement. By the time this, the fourth edition of *Living with haemophilia*, is published the first cure of either haemophilia A or B using the new techniques may have taken place, at least in animals.

Gene therapy has been approached slowly not just because of ethical or moral reasons, but because the doctors and scientists concerned want to ensure that it works and that it does no harm. It will be some years before it becomes routine for people diagnosed as having haemophilia and, even then, will probably neither be applicable nor available to everyone. So, whilst this edition is being revised with the dream of ultimate cure very much in mind, I hope that it will continue to help families cope with the everyday stresses and strains of living with haemophilia.

In developed nations the lives of people with haemophilia A and B have been made much more secure in recent years by spectacular improvements in the manufacture of blood products, and by the development of genetically engineered factor concentrates. The first of these, Recombinate made by Baxter, was licenced by the United States Food and Drug Administration on 10 December 1992. It was soon followed by Kogenate made by Miles, which was licenced on 25 February 1993. A third recombinant DNA concentrate will soon be available from the European manufacturer, Pharmacia.

These products represent years of brilliant research and development.

It is the profound hope of everyone who lives or works with haemophilia that they, and the eventual goal of permanent cure offered by gene therapy, will soon be available to those in developing countries. To this end, the World Federation of Hemophilia is now committed to the implementation of a strategic plan, the Decade Plan. The WFH President, Charles Carman, stated that the aim of this plan is to help bring effective therapy to the 80 per cent of the world's haemophilic population who presently receive little or no help in the management of their bleeding disorders.

I hope that *Living with haemophilia* will continue to help families to spring the trap of haemophilia, releasing children born with the disorder to the freedom of full and happy lives. In order to do this much of this edition has been completely rewritten. So, whilst as before it can be read as a complete book, it should now be easier to use it as a reference guide as well. In particular, there is a new introduction to the text for people unfamiliar with haemophilia. It takes them from the first days of coping with the diagnosis to what to expect as affected children grow to adulthood. I have also tried to simplify much of the technical material still further. Background Information Sheets have been included as guides for those readers wanting further knowledge, thus leaving the main text free for easier reading.

Newcastle upon Tyne   P.J.
September 1994

UNIVERSITY OF PLYMOUTH

Item No.   900 2974639

Date   1 7 APR 1997

Class No.   616·1572 Jon
Contl. No. 0192630300 X

LIBRARY SERVICES

The publishers are grateful to Alpha Therapeutics
and to Armour Pharmaceutical Company
Limited for their generous contributions
which have helped to limit the cost
of this publication.

Acknowledgements

Once again I am indebted to many friends and colleagues for their help and support during the preparation of this fourth English edition. In particular, the staff of the Newcastle Haemophilia Centre have continued to share their considerable expertise and enthusiasm for working with people with haemophilia and their families, who have themselves given so much of their wit and wisdom to the book.

Our clinical nurse specialist Maureen Fearns, physiotherapist Brenda Buzzard, and social workers Pat Latimer and Jeannie Fraser have all given invaluable advice. Linda McBride and Michelle Stenhouse were responsible for all secretarial help. Michelle not only put up with me as we battled through the everyday work involved in patient follow-up, but also typed the manuscript. Without these marvellous people there would be no book. Special mention must be again made of Dr Peter Hamilton, who finds the time to cope with all the administration involved in running haematology, as well as being committed to haemophilia. His unfailing support is deeply appreciated.

Outwith the Centre, John Gilroy has kindly provided the international list of aspirin-containing products. Dr Judith Goodship from the Department of Human Genetics has continued to work closely with us, and has kindly checked the updated text on inheritance, and Caroline Stockdale from the Dietetic Department has advised me on nutrition. Martyn Leates helped me revise the section on state benefit and employment.

Once again we are using photographs most kindly provided by Dr Tom Korn, Professor Franco Panicucci, Dr Nicole van den Bogaert, and Dr Françoise Verroust. Especial thanks are due to Stephen Tyler who gave permission for the reproduction of his photograph on the front cover. Stephen is an abseiller (and has severe haemophilia). The photo was taken at one of Dr Korn's annual camps for youngsters with haemophilia in Wales.

Within the Haemophilia Society David Watters, Graham Barker, and Mark Weaving have continued to encourage us to help produce attractive educational material for families some of which has appeared in the Society Bulletin, and some in literature published by the Bio Products Laboratory (BPL). Within the United States, the Medicine and Scientific Advisory Committee (MASAC) of the National Hemophilia Foundation has continued to provide up-to-date guidance for people with haemophilia and those responsible for their care. Their recent recommendations for prophylaxis are reproduced in the text with permission.

Without the encouragement and most generous financial help afforded by Alpha Therapeutic and the Armour Pharmaceutical Company the price of this edition would be out of the reach of many readers. I am again grateful to them and especially to Ian Marshall and Barry Barber, and to Lynn Wiesinger, John Sedor, and Alan Anderson. Thanks too for permission to reproduce the colour photographs provided by Boehringer Ingelheim.

This edition is the first to be produced with Oxford University Press. They have been terrific and my thanks are due to all the staff involved, for their expertise and unfailing patience.

In the first edition of *Living with haemophilia* in 1974 I wrote of my children's preferences for 'dinosaurs, dolphins, and mud'. Life has moved on and Mark (dinosaurs) is now a qualified surgeon, and married to Judith with a son, Jacob Blue. Emma (dolphins) is a civil servant and married to Ian with two boys, Adam and Benedict. Andrew (mud) is a qualified photographer, and is engaged to Rachel. They all continue to bring joy and happiness to Brigitte and me. We hope that our family's enthusiasm for life is reflected not only in this fourth edition of the book, but in the lives of all those with haemophilia and their relatives and friends.

Peter Jones
Newcastle upon Tyne, August 1994

Contents

Plates

Background

Some readers may, like the author, be of an age when a little background information may be of help in their understanding of the main text. The first insert is intended for parents and teachers and others who may, at one time or another, have responsibility for a boy with haemophilia. The other inserts describe features of the body which are relevant to haemophilia.

The English spelling of 'haemophilia' has been used throughout the text except where specific reference to the American spelling of 'hemophilia', for instance in reference to the World Federation of Hemophilia, is more appropriate. In describing people with haemophilia or the characteristics of their disorder, the adjective is 'haemophilic' (or 'hemophilic') and not 'haemophiliac'. For instance haemophilic bleeding, not haemophiliac bleeding.

All diagrams in the book have been prepared by the author, and all line drawings have been prepared by Doreen Lang and the artists of the Oxford University Press.

Dose schedules are being continually revised and new side-effects recognized. Oxford University Press makes no representation, express or implied, that the drug dosages in this book are correct. For these reasons the reader is strongly urged to check details with the staff of a haemophilia centre, or family doctor, and to consult the drug company's printed instructions before administering any of the drugs recommended in this book.

An introduction for new parents

Ten first thoughts for parents of a child diagnosed as having haemophilia

1. Your child is as beautiful, as cuddly, as much fun, and as smelly as he was before the diagnosis was made.
2. He will grow, develop, and thrive normally.
3. He is NOT condemned to a life of pain or handicap.
4. He will be able to play normally with other children and be as good, or as naughty, as they are.
5. He will NOT ruin the carpets by collapsing in pools of blood.
6. He will be able to go to normal school.
7. He has as much chance of becoming a top lawyer/doctor/politician as his friends.
8. He has as much chance of being Captain of one of his country's sports teams as his parents.
9. He will be able to travel and go on family or school holidays, at home or abroad.
10. His life expectancy is normal.

Initially, you will both feel hurt, and angry, and full of remorse, and these feelings will stay with you for a while. In the coming days, have a good cry (and shout at the cat) and then try to talk as much as you can to each other, and to your close relatives and friends. Make a list of your thoughts and fears so that you can talk them over together. Make a list of everything you are not sure about, however silly it may appear to be. Take this list with you when you go to see your child's doctor and go through it with him/her, point by point. Ask all the questions you want. At our Centre we call this the 'shopping list', because if you forget to bring it you are bound to forget something that has been worrying you!

Whatever you are told, a very natural reaction to the diagnosis is to think of the best ways you can to protect your child as he grows up. The following text is adapted from a series of articles I wrote for the Bulletin of the United Kingdom Haemophilia Society. I hope that it will help you come to terms with your fears.

Initially, all parents faced with a diagnosis of severe haemophilia, especially those with a haemophilic son as their firstborn, are fearful. They are unsure of how to handle their baby, and of how much they can play with him. They don't like going out for an evening. Neighbours who might otherwise babysit for them are frightened as well. The normal stresses and strains of bringing up a child are heightened by his haemophilia, and the fun of being a parent is clouded by the disorder.

Birth and the early years

Mothers with a family history of haemophilia sometimes worry about the effect of pregnancy, and especially delivery, on a haemophilic infant. There is no greater risk of miscarrying a haemophilic baby than a non-haemophilic baby. The womb and the bag of waters provide perfect protection. Nor is there any problem with a normal vaginal delivery which progresses gradually as the tissues of both mother and baby mould and slide over each other without injury.

During his first months the infant is protected like any other in his cradle. Parents who have been given the diagnosis already (from a cord blood assay of factor VIII or IX) should *never* be frightened of handling their new son, and of enjoying his development through smiling and sitting by himself, to his first attempts at speech. He does not need a padded cot, just an ordinary one, as he begins to shuffle and crawl and to pull himself up. He will enjoy a baby walker and a baby bouncer, and outings in an ordinary pram. He will like being cuddled and held up for the admiring scrutiny of various relatives and well-wishers. In other words, he should be handled just like any other child.

As he begins to toddle, like all toddlers he will fall over. He will collect bruises, especially on his forehead and legs. Most of these will be very superficial and they will not hurt him. This is the age at which parents begin to learn what to treat and what not to treat, so if in doubt always seek advice. If he hits his head very hard he must be seen by a doctor.

Sometimes parents ask about protective helmets and clothing. Personally, I dislike helmets especially when a child is old enough to walk steadily. Two of the families I have known have decked their sons out in fancy headgear; no harm has come to the rest. My view is that helmets worn regularly draw attention to the boy, labelling him as something different, and they can signal the first stages of an overprotective attitude that it is best to avoid. Helmets worn for cycling or other sports are, of course, sensible.

Other protective clothing is justified sometimes. If your child tends to bounce round the room on his knees, long trousers are sensible. Occasionally knees and/or elbows become the sites of frequent bruising and temporary protection with foam bandaging is required. Whenever a bandage of this sort is used for any length of time, at any age, it should always be taken off during periods of inactivity. If it is worn all the time it may cause muscle wasting, leaving the joint weak and unprotected.

Schooldays

There is great temptation to think of the haemophilic boy as a handicapped boy. He is not, but he can *become* handicapped if he doesn't receive the right treatment and fails to develop and mature normally. In order to ensure that this happens he must be allowed to explore the world and make mistakes, just like any other youngster.

Children barred from sports and physical education are at an immediate disadvantage. I have come to learn that the only sports that are really dangerous to the boy with haemophilia are boxing and rugby football, because they may result in head injury. Apart from these, I don't advise parents I meet to ban anything—except motor bikes when their sons feel like flexing their teenage muscles, and by then it's too late!

If a youngster is restricted and goes out and has an accident but is frightened to tell his parents, the resultant bleed will appear to be spontaneous, and his haemophilia more severe than it really is, and the restrictions reinforced. On the other hand, if he owns up and is punished for disobedience, the restriction is again enforced. He can't win, and trust within the family suffers. Let him learn his own limitations, and you will be rewarded as you watch him grow and thrive.

Let me end by telling two true stories. Every year boys from Europe with haemophilia spend their summer holidays at a seaside camp near Pisa. The camp is near the training ground for Olympic athletes, the boys participate in a wide range of sports—running, jumping, swimming, and throwing the javelin and discus among them. Yet bleeds are very rare—the boys are muscular and their muscles protect their joints. It has been estimated that only one accident in fifteen results in a bleed requiring treatment.

The second story is nearer home. One holiday my young son had a haemophilic friend to stay. In three days his friend had at least twelve accidents. He fell off his bike, tripped into a large rose bush, slipped in the stream, and stumbled out of a tree. My wife and I regarded him with awe—no bleeding! His parents hadn't restricted him and he knew how to relax as he fell. He did have a small shoulder bleed on his last morning, but that was successfully treated straight away with home therapy. When he left, my son said he was exhausted with having to keep up. That's one haemophilic youngster who will do well . . . if his parents can stand the pace!

You will find that all the points made in this short introduction are repeated and expanded elsewhere in the main text. This repetition is not a mistake. It is simply a way of helping to reinforce the ways of living with haemophilia in health and happiness.

PART ONE

Haemophilia

1

What is haemophilia?

Haemophilia is a disorder of blood clotting.

Normally when somebody injures themselves the blood clots in a few minutes and wound healing can begin. In haemophilia this does not happen. This is because one of the ingredients needed for making a blood clot does not work properly. This deficiency in the activity of the ingredient may be complete or partial. When the deficiency is complete the person is said to have severe haemophilia.

Even in severe haemophilia clotting is not prevented altogether. Instead it is much *slower* than normal. The result is a friable clot that cannot withstand the pressure of escaping blood. Body tissues involved become disrupted, and wounds do not heal up properly.

In severe haemophilia most bleeds are internal, into joints and muscles. With modern treatment given early there is often nothing to see, and no interference with everyday life.

The answer to the problems created by haemophilia is straight-forward. If the defective ingredient is replaced by an active ingredient, the blood clots normally.

This book is about how this is done.

The types of haemophilia

There are a number of blood clotting ingredients. Most are proteins. We make them continuously so that the body is always prepared for action should injury occur. As fresh ingredients are made and delivered to the bloodstream, so the older ingredients die and are either recycled or eliminated from the body. Normally, low levels of active ingredients circulate. Injury triggers a rapid increase in manufacture, and the bloodstream then delivers them to the site where they are needed. Here the clotting ingredients contain the damage and start the process of wound healing.

Most of the ingredients are called *factors*. There are 12 factors. One of them is called factor VIII, and another factor IX. It is the failure of the body to produce either normal factor VIII or normal factor IX that results in haemophilia.

When factor VIII is not normal (abnormal), the disorder is called
— **haemophilia A**.

When factor IX is abnormal, the disorder is called
— **haemophilia B**.

Haemophilia A is five times more common than haemophilia B. It is also referred to as factor VIII deficiency, as classical haemophilia, or sometimes just as 'haemophilia'. This last, generic term can be dangerous when it comes to giving treatment because factor VIII is no good to someone with haemophilia B.

Haemophilia B is also called factor IX deficiency or, especially in the United Kingdom where the first patient was identified, as Christmas disease. Christmas was the name of the patient.

Like factors VIII and IX, all the factors are identified by international agreement with Roman numerals. Sometimes newspapers refer to them incorrectly in the Arabic form as, for instance, factor 8 or factor 9.

Severity

When *all* a clotting factor is abnormal the resulting disorder is called severe. Conversely, when some of the factor is normal the disorder is only of moderate or even mild severity. In terms of figures someone with severe haemophilia A has a level of factor VIII in his blood of zero international units or, expressed as a percentage of average normal, 0 per cent. Someone with mild haemophilia A has over 0.05 international units per millilitre of blood (or > 5 per cent).

Knowing the *level* of the affected factor is important for three reasons:

- The level usually, but not always, indicates what to expect in terms of physical problems.
- The level indicates the sort of treatment that is likely to be successful.
- The level runs true to form in a family. So a baby with a family history of *mild* haemophilia will have *mild* haemophilia.

Inheritance

Haemophilia is an inherited disorder. This means that it can be passed down a family from one generation to the next. The instructions for making proteins such as factor VIII or factor IX are called *genes*. All the genes together form the blueprint for the life and the characteristics of an individual. There are 100 000 genes, collectively called the genome.

Half of a baby's genome comes from his father and half from his mother. The genes are carried in his father's sperm and in his mother's egg on structures called *chromosomes*. When sperm and egg fuse at the moment of conception the complete blueprint for the future individual is present.

Humans have 46 chromosomes. Two of these chromosomes determine the sex of an individual. They are called X and Y. Someone inheriting two X chromosomes (XX) will be female. Someone inheriting an X and a Y (XY) will be male.

The genes for haemophilia A and B are on an X chromosome. So they are said to be 'sex-linked'.

Because females have two X chromosomes they have two sets of instructions for making factors VIII and IX. If one set is faulty, the other set makes up for the deficit. That is why women do not usually have haemophilia. The inactive abnormal gene is masked by the normal activity of the other gene on the second X chromosome. The abnormality is hidden.

Someone with an abnormal gene like this is called a *carrier*. Carriers 'carry' the abnormal gene and can pass it on to their children.

Very rarely a women may inherit two abnormal factor VIII, or factor IX, genes, and have haemophilia. This can only happen when someone *with* haemophilia mates with a *carrier* of haemophilia.

Carriers may pass their haemophilia gene to their daughters or to their sons. The chances of this happening are 50:50.

If the X chromosome with the normal gene is in the fertilized egg the child *will not* inherit the haemophilia gene.

If the X chromosome with the abnormal gene is in the fertilized egg the child *will* inherit the haemophilia gene. If the child is female she will be a carrier like her mum. If the child is male he will have haemophilia because his Y chromosome does not have a duplicate set of instructions for making factor VIII, or factor IX.

A *father with haemophilia* may fertilize an egg (X) with a sperm containing either an X chromosome or a Y chromosome. His X chromosome results in a daughter (XX). Because his X chromosome bears the abnormal gene all his daughters *must* be carriers of haemophilia.

His Y chromosome results in a son (XY). Because his Y chromosome is normal all his sons must be normal. They cannot inherit a haemophilia gene and, therefore, cannot pass haemophilia to any of their children. The line of inheritance ends.

About one third of all genetic disorders seem to come out of the blue. There is no family history. This is because our genetic make-up undergoes changes, and one of these changes can result in haemophilia. A change like this is called a mutation. In the family history of someone with haemophilia there is no way of telling (except occasionally with specialized testing) the timing of this mutation. It may have just happened to the fertilized egg, or it may have happened generations beforehand and have been unknowingly passed down the female line where, of course, it would have been masked by normal X chromosomes.

So, in summary:

- Each daughter of a carrier mother has a 50 : 50 chance of being a carrier too.
- Each son of a carrier mother has a 50 : 50 chance of inheriting haemophilia.
- Each daughter of a father with haemophilia must be a carrier.
- Each son of a father with haemophilia must be normal, and cannot pass haemophilia to his children.

How do I know when treatment is necessary?

It takes a little time to get used to living with a child who has just been diagnosed as having haemophilia. The label haemophilia constantly gets in the way. It is easy to think of nothing else, and to start to imagine a disrupted and difficult future for him and the family. In fact nasty bleeds are *very* rare in early childhood and there is nearly always plenty of time to learn how to cope with everyday knocks and bruising before more serious bleeds begin.

Diagnosis from cord blood or in the newborn period is followed by several quiet months in which haemophilia produces few, if any, visible signs. As the boy with severe haemophilia becomes more mobile bruising appears. Although unsightly these bruises do not hurt him. Unless there has been a big knock they are superficial. Try moving them around in the skin with your fingers. They feel nubbly and move easily over underlying tissues. If you can do this no treatment is necessary. If you cannot, seek help. Fixed bruises suggest bleeding into deep tissues and probably need treatment.

Bleeds over or very near joints should also be looked at by someone with experience of haemophilia, unless they are very small. Bleeds into joints, causing swelling of the joint, are uncommon before the age of 2 years but they do occur. When they do early treatment is very important if long-term damage is to be avoided. This is why some doctors are now recommending that prophylaxis, in which regular shots of factor VIII or

IX are given, be started in the second year of life if the boy's haemophilia is severe.

Apart from joint or deep muscle bleeds the one time that advice MUST ALWAYS be sought is in the case of head injury. All toddlers fall and bump their heads, and it can be very difficult to decide whether or not a bruised head should be seen. My advice is that all knocks to the head that would worry the parent of any child must be seen and probably treated. Included here are direct falls from a height, falls on to concrete, blows as a result of running directly into a wall or vehicle, and falls from moving tricycles and bikes. The golden rule is 'if in doubt, treat'—this is one occasion when it *is* better to be safe than sorry!

Tests and immunizations

Screening tests on the newborn, like the Guthrie test for phenyl-ketonuria, often involve taking a small specimen of blood from a heel prick. This only hurts for an instant and does no harm to a baby with haemophilia. What should not happen, except in real emergency, is an attempt to take blood from a blood vessel in the neck (an external jugular vein) or the groin (a femoral vein) in order to make a diagnosis of haemophilia. These vessels are relatively large and bruising around them can be very dangerous. If blood is taken from a neck or groin vein make sure pressure is kept on the site for at least five minutes, and that the area is then checked regularly for several hours. Untoward swelling requires further pressure and sometimes special treatment. If the diagnosis has not been made on a specimen from the umbilical cord it is best to wait until a good vein in the hand, wrist or crease of the elbow, or on the top of the foot becomes available. Veins show up better as babies grow and start to move around more. Unless a baby is very fat access to one of these peripheral veins should be relatively easy before his tenth month of life.

Immunizations are fine before this time. The injections are of small volume and, provided finger pressure is kept over the injection site for five minutes, do not cause untoward bruising. I have only had to treat two children for bleeding after immunization in over 30 years of haemophilia practice.

Other than immunization *all* intramuscular injections are banned in people with haemophilia. Medicines must be given via another route, usually intravenously (into a vein). This is because medicines, especially antibiotics, are often of large volume and can easily provoke extensive bleeding into muscles.

Treatment

Nowadays treatment is easy, at least in developed countries where access to quality blood products is available. More about how to cope in the developing world will be found in Chapter 13.

Treatment consists of replacing the missing clotting activity of factor VIII or IX.

In the case of haemophilia A this can be done either by using a product made from blood plasma provided by human donors or, in some cases by pigs, or by using a factor VIII preparation made synthetically by bioengineering. In all cases the product will have undergone stringent testing and measures to remove viral contamination.

In the case of haemophilia B the preparation available is likely to be a factor IX concentrate made from human plasma, although a synthetic product is being developed at present.

These concentrated forms of factors VIII and IX are expensive and must be used wisely. They are *very* effective and their introduction has revolutionized the management of haemophilia. Before they were available in large quantities treatment was with unrefined blood plasma in the form of fresh frozen plasma or cryoprecipitate, both of which had to be stored in a deep freeze. These simple products are very effective but are less safe from viral contamination than the concentrates, and are harder to store and administer.

Prospects for life

With modern treatment the child with severe haemophilia can expect to live a normal lifespan. The quality of his life should be no different from that of his friends without haemophilia. Careful on-demand therapy or regular prophylaxis with a virally safe clotting factor product should guard him from arthritis in later life. He should be safe from the viruses that cause hepatitis and infection with the human immunodeficiency virus that causes AIDS.

If he has mild haemophilia his major task will be to remember the fact! If he forgets he may get a sharp reminder after a visit to the dentist or after what his surgeon thought would be a minor operation.

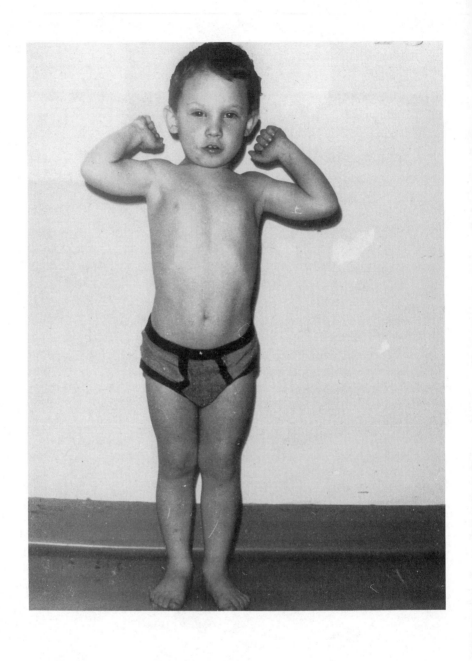

2

Growing up with haemophilia

Haemophilia is not a disease. Children are born with it. If the parents know that there is a family history of haemophilia their sons can be diagnosed in the womb, or at birth. When there is no family history severe haemophilia usually presents itself in the first year of life. Less severe forms may come to light at any time. People with very mild haemophilia may only be diagnosed when they are well into adult life and have bled for longer than usual after an operation or dental extraction.

Parents told that a son has severe haemophilia feel overwhelmed. The diagnosis is bad enough, but having to come to terms with all the medical information and jargon is something else! When you are told the diagnosis you need to be assured that your child is secure. You need to know who to contact and how to contact them at any time if you are worried. Until now your major health contacts will have been your family doctor and health visitor or midwife, and perhaps an obstetrician and a paediatrician. You now need someone with experience of looking after haemophilia.

Many countries have networks of haemophilia centres, and the doctor who made the diagnosis on your son should know the nearest centre to your home. If there are any problems with this you can seek help directly by getting in touch with the nearest haemophilia society branch or chapter. Health information services at your local library may have the address of at least the national headquarters. If you are stuck, contact either the service in another country for information, or the headquarters of the World Federation of Hemophilia.

The current address is:

> The World Federation of Hemophilia
> 1301 Greene Avenue
> Suite 500
> Montreal
> Quebec
> Canada, H3Z 2B2
> Tel: (1)–514–933 7944
> Fax: (1)–514–933 8916

Once you know how to get help when you need it, *relax*. Some of the joys (and the trials) of bringing up a boy with haemophilia have already been mentioned in the 'Introduction for new parents.' The information given in this chapter is taken from articles I originally wrote for the UK Haemophilia Society. It should add further relevant and practical information to those first thoughts.

When you have read this chapter think about meeting up with other families with a child with haemophilia. Sharing a problem is really a problem halved! Think also about reading some of the literature about haemophilia published in America by Laureen Kelly. Two of her publications are *The parent exchange newsletter* and *Factor friends*. Her address is 6 Marshall Street, Medford, Massachusetts 02155, USA.

Babies

Babies are fun! When a baby with haemophilia is born he is just as much fun as any other baby. Nothing in his childhood needs to be different.

It is terribly important to realize this. The 'START-LINE' for any baby must be **normality**. This is true of any disorder, because if the normal things are not recognized and nurtured they will not thrive. When this happens the disorder can eventually take precedence over everything else, smothering the child and his family and eventually leading to the self-recognition of haemophilia as a handicap. A diagnosis of haemophilia is *not* a handicap; it can only become a handicap if the family or their medical advisers allow it to become one.

> 100 000 genes carry the instructions to bring a baby to life.
>
> The only difference between a baby with haemophilia and a baby without haemophilia is that *one gene* does not work properly in the haemophilic child.
>
> I think that the other 99 999 genes deserve priority, don't you?

Pregnancy, birth, and diagnosis

When there is a family history and the prospective mother knows from this that she may be a carrier, she and her partner should talk to doctors with experience of up-to-date haemophilia care. Any decision about having a child who may have haemophilia always lies with the couple themselves. However, doctors can help in the diagnosis of carriership and sometimes, if the couple want it, the diagnosis of haemophilia in the womb.

If a twin pregnancy is diagnosed what are the chances of both babies being affected? Identical twin boys will of course both have haemophilia if they have inherited the gene. Non-identical twin boys may or may not both have the condition, it depends on which of the mother's X chromosomes each boy has inherited.

When a couple decide that they want the pregnancy to go to term, arrangements can often be made for diagnosis to be carried out on a specimen of blood taken from the umbilical cord immediately after birth. This hurts neither the baby nor his mother. It is important that good facilities are available both within the maternity hospital for taking the specimens, and within the nearest haemophilia centre so that the specimens can be tested very quickly. Any delay may give a false result. Unless there is an emergency blood for diagnosis should not be taken from veins in the neck or groin. There is no need. If the opportunity to take a cord blood specimen has been missed the diagnosis can wait until the boy is older and better venous access has developed.

Haemophilic babies are born perfectly normally and are not in any danger during normal vaginal delivery. This is because the smooth, flowing nature of delivery, during which the tissues of both the baby's head and the birth canal are gradually moulded, does not present the baby with the sort of challenge that causes haemophilic bleeding. Caesarian section is only indicated when there are particular diffi- culties during the delivery, for instance a potentially difficult forceps procedure. On the sixth day of life all babies have blood taken from a heel for the Guthrie test for phenylketonuria. The skin prick is small, and I have never known problems even when the infant has severe haemophilia.

When there is a family history the parents' attitudes to a newborn baby with haemophilia will obviously be coloured by the family experiences. If these have been bad with, for instance, particularly nasty bleeds or experiences of surgery going wrong, or premature death of relatives, then the couple will be more fearful for their child than they would be in a family where all the experience is good. Research and development in haemophilia is so rapid that it is very important for couples in this position to get up-to-date information before their child is born, so they can act on fact and not fantasy. Old medical textbooks from the local library are NOT recommended.

When there is a possibility, because of a family history, that an infant may have haemophilia his parents should remember to tell his doctors. This is especially important if any operation, for instance circumcision, is contemplated in the early months of life.

When there is no family history the diagnosis of haemophilia will not be made at birth. It will probably follow either the appearance of bruising when the baby is around one-year-old and beginning to be more mobile, or perhaps prolonged bleeding following injury to the mouth. Typically, the delicate piece of tissue between the top lip and the gum is injured. In normal children bleeding stops quickly and stays stopped. In a child with severe haemophilia it commonly restarts again because healing is easily disturbed by movement during eating or chattering.

Unfortunately, because haemophilia is so rare (around 1 in 10 000 people) its diagnosis may be delayed for some time and doctors may look for other causes of bruising including, in some cases, non-accidental injury. This is awful for any parent, but unfortunately reflects our society and the fact that babies who have been injured intentionally form a normal part of the life of any hospital children's department. Clotting tests soon lead to the right diagnosis.

At the toddler stage inquisitive people with nothing better to do may start remarking on your son's bruises. Armed with the diagnosis, insinuations about people who hit their children can now be treated with the contempt they deserve. Most mums of haemophilic boys come up against malicious suggestions at some stage. This is one of the problems of having affected children who look perfectly normal most of the time. It is something to learn to live with. When faced with such people either try explaining the reasons sensibly, or ignore them. Or, if the provocation really is too great, 'spit in their eye' and forget about it!

Families who have had other children before the son with haemophilia already have experience of the normal growth and development of children, with all the successes and setbacks usual as kids grow up. They can, therefore, tell the difference between the trials and tribulations of everyday life and the problems caused by haemophilia. The couple with a first-born baby with haemophilia have a lot to learn. Life can be especially difficult if they have been blessed with a hyperactive youngster whose only intention is to take command of the family, day and night.

Things that need doing once you are home with your baby

First and foremost is the need to have a good family doctor and to have access to a good haemophilia centre. The family doctor does not need to be an expert in the condition, but he or she will give all the usual support needed by a couple and their children. The haemophilia centre staff work with the family doctor whilst providing the specialist care for the haemophilia. You should also join the Haemophilia Society.

Make sure that you understand your son's diagnosis and are given a card filled in with the details needed for treatment. In the UK this is called a 'Special Medical Card for Haemorrhagic States'. In particular, it is important that his factor VIII or IX *level* is recorded because on this depends the specific treatment of his disorder. For instance, if he has a factor VIII level of over 5 per cent he may be able to be treated with desmopressin rather than with a blood product. Throughout his childhood his doctor will routinely test him for inhibitors to factor VIII or IX. The discovery of an inhibitor can mean that a change in treatment is needed.

Like any child, the boy with haemophilia needs to grow up in a secure environment. Injuries are more likely to occur in the cramped, crowded conditions of inner cities with steep stairs and access only to busy streets, than in more spacious houses with gardens. Help can often be given in re-housing by staff of the haemophilia centre. Every family with a child with haemophilia should have a telephone in the house, and again haemophilia centre staff can often help persuade local authorities to provide telephones for families who could not otherwise afford them.

If both parents go out to work and have to leave their boy in a crèche or nursery school, they should make sure that the supervisor knows about the bleeding disorder. They should stress that the boy be free to play with other children normally, and not be restricted. The supervisor should be given a telephone number where they can contact a parent. Mobile telephones and pagers are very useful here; some companies provide them free of charge to families with haemophilia. Babysitters can also be armed with knowledge of where a couple are. All parents need to escape the demands of their offspring at times, and the parents of a haemophilic child are no exception. They need to get out to the cinema, local pub, or restaurant just like anybody else.

One of the family doctor's roles is to ensure that all children are immunized against the full range of infectious diseases. When an injection is needed, that too is perfectly safe in haemophilia because the volume of the injection is small. My advice is to keep pressure over the injection site for four or five minutes to prevent any untoward bruising. In the case of BCG, for the prevention of tuberculosis, there is no need to put pressure on. In addition to the usual immunizations in childhood all children with haemophilia should be vaccinated against hepatitis B. There is also a vaccine against hepatitis A. No vaccine yet exists for hepatitis C.

Other than the injections needed for immunization, all other intramuscular injections are **banned** in haemophilia. Medicines are given by vein instead. **Remember that no child should be given aspirin nowadays**. This is especially important in haemophilia because the drug interferes with the stickiness of the blood platelets, and adds to problems with bleeding.

It does not matter how babies with haemophilia are fed, although my general advice to all mothers is always to try to breast rather than bottle feed. Weaning is normal. Nor is there any reason for handling babies with haemophilia differently than any other baby.

Some authorities on haemophilia used to suggest that padding should be applied to furniture and to the baby's cot. This is a throwback to the days of the Russian prince Alexei for whom the trees were padded in the park! As with most things in life, a little common sense goes a long way and, provided heavy ornaments and Aunt Edith's favourite vase are kept out of reach, all will be well. It is sensible, as with all children, to have a gate fitted to prevent access to steep stairs, to avoid placing mats on highly polished floors, and to check toys to see that they are not likely to present sharp or jagged edges after the usual bashing. A child's safety seat or harness should be fitted in the family car and used, even for short journeys.

The boy with haemophilia should be cuddled and played with, and swung, and enjoyed by all the family, relatives, and friends just like any other child. There is no need to make any alterations to pushchairs or prams. He can have a baby bouncer and a baby walker, but make sure the walker cannot tip up easily as he becomes more adventurous.

When I was young clothing for babies was dreadful. Everybody, whether boy or girl, seemed to be attired in white dresses with plenty of lace. Nowadays clothes for youngsters are super, bright, and colourful, warm in winter and cool in summer. That's fine but make sure that, whatever its colour, his footwear is of good quality as he grows up.

Early bumps and bangs

As the child becomes mobile and starts to explore the world and bump into pieces of furniture, the first evidence of easy bruising appears. Dungarees will help to protect his knees on hard surfaces as he crawls. At the age of 11 months or so he will start pulling himself up and collapsing with a bump on his behind, and will then start creating havoc with prized possessions kept on low tables. He will also have discovered how to negotiate stairs, and will invariably be found in the last place in which his distracted parents look. Bruising may appear on the arms, legs, hips, and buttocks, and sometimes on the trunk, but he is well padded with fat and this gives him natural protection. The bruises are almost always superficial and rarely painful or dangerous.

An exception to this is bruising around the head. All small boys bang their heads as they start to toddle and lose their balance, but some seem to be especially prone to this type of injury. For them, protection may be required. Cycle and sports shops nowadays stock a range of colourful helmets for children. Choose one that can be adjusted to fit snugly and not come off in a fall. This sort of protection is needed only very rarely for everyday use, other than for sports which may result in knocks to the head. Once the child is steady on his feet discard the helmet, except for its use in sports.

Some common questions

Three particular questions that are sometimes raised during infancy concern circumcision, teething, and anal fissures.

Circumcision should not be performed unless there is a real necessity. In medical terms this means good evidence that a tight foreskin is obstructing the flow of urine or sometimes becomes inflamed, rather than that the foreskin doesn't retract—this is usual in many boys until they are around seven-years-old. If there is a medical reason, or there are irrevocable religious reasons, then the procedure must be carried out very carefully in an operating theatre with full haemophilia protection. This will mean the baby and his mother or father staying in hospital for a few days.

Babies with even severe haemophilia have no more than the usual trouble when they teethe. Later, when they start to lose their baby teeth all is normal too. The teeth are pushed out very slowly with only a minimal amount of bleeding that does not require treatment.

A lot of babies develop small tears around the anus after passing hard stools. A tear like this is called an 'anal fissure' and it distresses the baby because it is painful to him during defecation. In haemophilia it can also lead to a little more bleeding than usual. Fissures heal up easily if stools are softened by, for instance, adding more sugar or fruit juice to the diet of a baby being artificially fed. Sometimes a little local anaesthetic cream can help. I have never known a baby with haemophilia and an anal fissure require specific haemophilia treatment.

Follow-up

Other than the need for regular follow-up at the haemophilia centre, most families need to spend very little time visiting a hospital in the early years. When they do, it is extremely important that fathers make every effort to attend as well. In our society it is the mothers who are often expected to take their children to hospital for follow-up and to cater for their bleeding episodes. In the old days this attitude used to lead to haemophilic children growing up without real recognition of their dads and sometimes with over-dependence on the love and care of their mothers.

Main lesson of the day for this age group:
ENJOY HIM!

Checklist

Enrol with a good family doctor
Make contact with your nearest haemophilia centre
Join the Haemophilia Society
Know everything about the diagnosis, including the clotting factor level
Have a haemophilia card issued for your son
Learn about the treatment he may need
Let everybody in the family know about him
Install a telephone at home and know how to get help when needed
Go to the baby clinics
Have him immunized
Involve his dad, and his brothers and sisters, in his care

Toddlers and pre-school children

Being able to watch young children explore their world provides one of the great rewards of life. Toddlers cheerfully follow a finger—usually into all sorts of potential trouble. Those with haemophilia are no different. They need the stimulus of exploration in order to develop normally. And they need that stimulus in a secure and loving environment. That means a family without fear. Children constantly sense atmosphere and if their parents are afraid of haemophilia, they will be too.

Coming to terms with haemophilia takes time. Occasionally you will need to cry—to mourn the 'normal' child you wanted. That's fine. Let it go and then go back to your child, and to the normality of the fun and excitement of watching a new life take shape. You will know you are through the worst when you look at your child and see the smile and not the haemophilia.

Three things can go wrong in those wonderful pre-school years when all parents relive their early days through the lives of their children. These are accidents, bleeds, and over-protection.

Accidents

Accidents, of course, happen to any child. Many can be avoided. Lock up medicines and household cleaners. Ensure the garden is secure and that dashes to freedom amongst the traffic are impossible. Keep matches out of reach, and use a fireguard. Common sense for any child.

Whatever you do, help your child explore his world early. Don't hold back because of his haemophilia. Take him to the swimming pool, the park, and the playground. As he grows he needs to learn about play and about how to relate to other children. As he does he will also learn how to control his haemophilia, rather than let his haemophilia control him! The list in the box on p. 22 emphasizes how normal the life of a boy with haemophilia should be.

As babies with haemophilia start to become more mobile some parents like to reassure themselves by padding the knees of long trousers, and sometimes sleeves to protect elbows from bruising, but I have yet to see a bleed into a baby's joint as a result of bangs at this age. Bruises are common but are usually superficial and not painful. Whatever you do, don't bandage the joints themselves. That leads to muscle wasting and instability. All babies fall down but they are well padded anyway. Nature provides them with that nice protective fatty layer that we all spend the rest of our lives trying to get rid of.

The only area which is not protected in this way is the head, and one of the worries of growing up with active haemophilic youngsters is head injury. All toddlers bang their heads either by falling or walking into things. However, it is very rare for the ordinary knocks of everyday life to result in bleeding inside the head. I always encourage parents to have a child who has had a BIG bang seen by centre staff because early treatment will prevent problems. Certainly if a bang to the head has been particularly severe (for instance a hard fall on to concrete, or a child running into the side of a car) he should receive treatment. In these circumstances it is sometimes wise for him to be admitted for a period of observation as well. It must be stressed that injuries leading to this are very rare indeed, and that is one of the reasons I do not recommend helmets for everyday play. The other reason is because the over-use of protective clothing in ordinary activities encourages the development of awareness of handicap.

Bleeds

Bleeds at this age usually occur as a result of a bang or a fall. Superficial bruising is the most common form of bleed. Abrasions and small cuts cause no more trouble than in anyone else, they just need a sticky plaster. Joint bleeds are uncommon in the first three years of life, but they can occur. If you are in doubt seek advice straightaway.

Admission to hospital for treatment is nowadays extremely rare. If bleeds do require treatment it is given as an out-patient until the family are ready to start home therapy. Problems can arise here because the doctor on call on a particular day may be relatively inexperienced in giving intravenous therapy to small children. Parents should take advice from the staff of their haemophilia centre about the best way of giving an individual child his treatment. In general, children should never be taken away from their mother or father. It is far easier to treat a child when he is sitting on a parent's lap.

A small-vein set (a fine needle attached to a length of polythene tubing, which is then attached to the syringe) is far and away easier to use than a rigid needle attached directly to a syringe. On no occasion should the same doctor or nurse attempt more than three venepunctures. If he or she fails three times success is unlikely and a more experienced person should be called. In any case children who have to undergo more than one venepuncture should be rewarded in some way. There is nothing like being out of pocket to a toddler to improve the doctor's venepuncture technique!

Two other points. Firstly, parents should know the positions of the best veins to use in their child. Secondly, all this will sound very threatening to new parents, but they can be assured that they will soon get used to the procedures involved and become very skilled at handling their child so that he can have treatment promptly. It is always remarkable to see how children adapt to treatment which, after a while, does not even hurt. It is also remarkable how quickly parents become able to relax and cope with bleeds and with treatment, no matter how scared they are initially of needles.

Pain associated with haemophilia is rare in the pre-school child. When it occurs the best medicine is paracetamol, as in any child. Remember always to keep paracetamol preparations well out of reach;

overdosage can be lethal. Remember, too that *medicines contain-ing aspirin are not suitable for children and especially for any-one with haemophilia*, because aspirin makes the bleeding disorder worse.

Over-protection

The third thing that can go wrong occurs as a result of fear, especially when parents feel themselves isolated. It is a natural, loving reaction to protect one's children. Over-protection happens when concern about the haemophilia leads to attempts to avoid all the everyday mishaps of normal life. I have come across families too frightened to let their children out of their homes, parents who only let their children play with soft toys, and parents who prefer friends not to call because they are so frightened that a childhood scrap might result in a devastating bleed. The results of this sort of over-protection start to become apparent very early in life and are at their most severe in late adolescence and adulthood. Failure to come to terms with haemophilia leads to isolation, loneliness, and a lessening of opportunities for marriage and a decent career. One of the best ways to avoid over-protection is to encourage activity with family and friends from an early age. Parents should not be afraid of sharing their haemophilic son. All children with haemophilia should be encouraged to participate in activities and sports with others.

The more the boy mixes with others the better. They will get used to his disorder and learn not to be frightened by it, and he will be able to explore his relationship with them in a normal way. If this approach is cultivated early the family will be well on the road to avoiding over-protection. The over-protected boy grows up crippled more from the misguided love of his parents than from his bleeding disorder, and his normal brothers and sisters take second place and suffer too.

The boy's father is a very important person in this context. There is a tendency for the child with haemophilia to be mum's responsibility—a tendency often extended to the hospital where the majority of treatment is given by female nurses and physiotherapists. The father should play with his child as he would play with a normal boy. Romps round the bed, splashing together in the bath, hide and seek, fun fights, and swinging games, and opportunities to respond to male relatives are part of growing up. Hospital visits should fit into this pattern, and the father should be prepared to take his son to the clinic when he can. The sharing of childhood with the boy by both parents will help them, too. Fathers are sometimes reluctant to admit their boy has a bleeding disorder, and initially feel shy and strange in a medical environment more familiar to their wives. An early effort to come to terms with these very natural feelings will be amply rewarded in the future. The family will be strengthened by their mutual experience in the bad times as well as the good.

What's normal	What's abnormal
Bikes	Factor VIII or IX clotting activity
Clothing	
Going out with other people	
Growth	
Holidays abroad	
Immunizations	
Intelligence	
Jumping	
Kissing	
Life expectancy	
Physical development	
'Pig-outs' at McDonalds	
Play	
Playschool	
Potty training	
Running	
Swimming	
Toys	
Tricycles	

Introducing sport

The first sport which most children enjoy is **swimming**. Swimming has just been voted top sport for someone with haemophilia in an international survey of the views of doctors by the World Federation of Hemophilia. Toddlers love going to the pool, and nowadays most neighbourhoods have excellent facilities for parents to take their children to play in the water. Playing in warm water allows exercise of all joints and muscle groups in the body without gravity, and this can be particularly beneficial after a joint bleed. In addition, swimming helps coordination and the development of strong muscles which protect the joints. This in turn lessens the risk of the development of haemophilic arthritis in later life.

No child should grow into adolescence without being able to swim.

The activity may also help the mother. One of the effects of haemophilia on her used to be a sore back induced by lifting and carrying her haemophilic son during bleeding episodes. With home therapy this should be unusual nowadays, but the presence of clotting factor inhibitors or a particularly bad run of bleeds will put additional strain on her. Regular exercise in the water will reduce the ache by strengthening her muscles.

Some thoughts about restriction and punishment
Restriction

Undue restriction of the growing boy's activities can have remarkable long-term effects. At some stage of his life, notably during the years of early exploration with his peers or in his teens, the boy will reject restriction. This rejection is, by itself, natural. It is a part of normal development without which the adult cannot be a complete and independent person. If a chronic disorder like haemophilia is used by those in authority—be they parents, doctors, or teachers—as the reason for restriction, a familiar pattern emerges. The youngster begins to hide his rejection, appearing to adopt the role of acceptance expected of him. If accidents occur—as they are bound to—they will, when possible, be hidden. Resultant bleeds will be recorded as spontaneous, and the very 'spontaneity' of the haemorrhage will serve to confirm the 'severity' of the disorder, thus reinforcing the parents' (and the doctor's) original concept of restraint.

Paradoxically, the discovery of an accident leading to a bleed because of a boy's failure to conform to the wishes of his advisers also reinforces the restriction. The youngster cannot win.

Restriction has other consequences. Boys forbidden to explore their potential in sport and manual dexterity receive no training in skill, and the acquisition of skill in movement or in the use of tools protects from, rather than provokes, haemorrhage. Most boys will take sensible advice if they can see that the reasoning for a restriction is sound—for instance, that boxing is likely to result in head injury. It is the constant repetition of theoretical restrictive advice which will result in that sense of failure hinted at in some older men with haemophilia. The boy brought up to view the world as a hostile place and to recognize bleeds as the inevitable consequence of not doing what he is told is very much on his own. Not only does he have no one to turn to when he gets into real trouble, but he misses that sensible rapport and monitoring that will eventually help him to build a sense of responsibility, and a pride in his body and his achievements.

Restriction and its denial can also, if not recognized and handled with

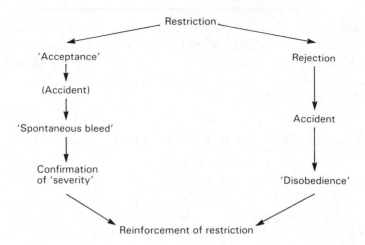

Fig. 2.1. Restrictive practices in haemophilia. Unquestioning acceptance is likely to result in dependency in adulthood. Rejection and hiding of accidents leaves a boy very much on his own—he grows without sensible guidance.

care by the doctor, result in unnecessary emphasis on medical treatment. Following a run of spurious 'spontaneous' bleeds, prophylaxis may be substituted for on-demand therapy, or admission to hospital for orthopaedic treatment be advocated.

Punishment

Few parents of haemophilic boys admit to the smacking of bottoms as a punishment for naughtiness! My own opinion is that, as for other children, no harm will come from the occasional smack, providing it is used only as a last resort and never when the parent is angry, and is always administered with the flat of the hand on the child's bottom. A better punishment is to send the boy to his bedroom for a while—though to be really effective this procedure has to be combined with the loss of a favourite television or video programme or computer game. In some families, punishment like this is graced with the name 'time out'. From an early age, the child is trained to understand that unacceptable behaviour will inevitably result in his exclusion from whatever the family are doing, or what he would like to be doing. The period of exclusion varies with the gravity of his offence!

It is very important not to scold a child simply because he has had a bleed. This may make him reluctant to admit to bleeding in its early stages on future occasions. Remember that bleeds are not necessarily due to recognizable injury—they may really be spontaneous.

Advice to others

As a child grows he should be allowed out to play with other children and eventually go to playschool. The advice that follows is that I give to all parents.

Firstly, anyone with a responsibility of looking after a child with haemophilia should know that **all usual first aid measures apply**. Children with haemophilia bleed no faster than other children and there is always time to get help. This is one of the points I always make in letters to teachers. I write these to give to the parents to take to the school themselves. In this way teachers can see that the parents themselves are in control and know all about their son's haemophilia.

When writing to a playschool emphasis should be put on the fact that John can do everything from riding tricycles to bashing other children in the sand pit, and that no special restrictions should be enforced. He should be left to explore his new world in the same way as the other children. Just like them, he will learn to climb and use the slide—the right way up and upside down—and to swing and jump. There are some funny myths about haemophilia, and I think it also worthwhile reassuring teachers that the condition presents no threat to other children. They will be reassured to learn that the great majority of bleeds are internal anyway.

Other than teachers, babysitters and, of course, grandparents and other relatives, the only people who really need to know about a child's haemophilia are the family and other doctors who may be consulted, and the family dentist. Going to the dentist should be a regular routine with an older brother or sister, or mother or father, from an early age. Although nothing active needs to be done the young child learns not to be afraid, and he has an introduction to dental hygiene which will be of great importance to him in later life. The doctor at your haemophilia centre will prepare letters about your son for his medical and dental colleagues. Whatever you do, always be open and honest about your child's haemophilia. The more it is talked about the easier life will be. Remember, too, that all children like attention and rewards. Both should be distributed evenly throughout a family. Haemophilia should not be an open passport to the sweetie tin! If you find yourself tempted, remember that fat children with haemophilia fare worse than thin children with haemophilia. Extra weight puts more stress on joints, and fat children are clumsier and more likely to have accidents than thin ones.

Parents who manage to avoid the pitfalls, and look to the normality of their child rather than to his haemophilia, will be amply rewarded.

Main lesson of the day for this age group:
LET HIM EXPLORE!

Checklist

Let him play with other children
Don't be frightened of going out and leaving him with a babysitter
Let him explore
Enter him for playschool
Buy him a tricycle
Give time to his brothers and sisters
Both parents should be involved
Compare notes with other parents
Trust your common sense

Going to school

Every family album has a first-day-at-big-school picture in it. Tom, aged 4, in school uniform big enough for him to grow into. Beautifully pressed short trousers. Smart school tie. Socks. Clean shoes. It doesn't last long.

First days away from home are always a wrench. Hopefully, by the time of the start of his formal education the boy with haemophilia will have had plenty of opportunity to play and mix with other children. Nursery or playschool should have defused his mother's initial fears of letting other people look after her child—and her child's bleeding disorder. This earlier informal education makes the start of formal schooling easier. Parents will have confidence that their son is able to cope for short periods of time without them. They will know that he can be trusted to the care of others. The boy is used to the discipline of structured play with other children, and will have survived more than one scrap.

Before that first day his parents should have made sure that the teachers are both aware of his haemophilia, and are comfortable with the diagnosis. This can only happen if they have up-to-date information from the family and from the haemophilia centre. All teachers are used to coping with a wide range of disorders at school. For instance, asthma, diabetes, and epilepsy are all more common than haemophilia. It really takes an awful lot to faze the average teacher! But they do need that up-to-date briefing, and they need to know what to do if Tom presents them with a problem.

The best way I have found to do this (as with the playschool age group) is to give a personal and more detailed letter about the child to his parents (see p. 322). Again, it is they who take the letter when they go to see the teachers. The message is clear. They, the parents, are in control of all decisions taken about their son.

Before looking at the specifics of a school letter it is worth while emphasizing once again some of the features of normal development. Play in the early school years is not usually formal or competitive, and the haemophilic boy should be allowed to participate in all activities. This is the time during which the child's world begins to expand beyond his family and immediate neighbours, and in that exploration he begins to learn his gifts and limitations. Too much emphasis on restriction will harm him; he must learn the limits over which he cannot go without injury for himself. Provided that his teachers know about the bleeding disorder and how to help when it is really needed, common sense should be allowed to prevail.

The years between eight and ten are the years of the first gang or club, and a little restraint may become necessary when 'dares' may start to overcome natural caution. All boys climb trees, with or without their parents' approval. If a treehouse appears in the neighbourhood, one of his parents should check it over for safety. Friends should know that young Tom bruises easily, but they will not take kindly to his observation that, whilst they are not allowed to bash him, he can thump them with impunity.

Most children like to try carpentry, metalwork, or help with car maintenance at some stage. This should be encouraged, as should any activity that might become a serious hobby. It is sensible to supervise early attempts, and especially to teach the correct use of machinery and tools. Many men with haemophilia are very clever with their hands and can take pride in their workmanship over the years. If bleeds occur, they are usually minor and readily treated. Recurrent bleeds traced to a cause connected with a hobby may be prevented by a simple change in technique, better equipment, or special clothing or support.

When competitive sports start, certain restrictions do become necessary. These will depend on the severity of the particular bleeding disorder in the individual child. In general, body contact sports are to be avoided. These include boxing and rugby football. Harm seldom comes from a boy kicking a ball about in play, but I used to suggest that competitive football be banned. Experience has shown that I was wrong. I now feel firmly that the child keen to participate with his pals on the soccer pitch should not be discouraged. Injuries are rare and are almost always confined to minor bruising. The same goes for most other sports with the exception of boxing and rugby football. Of course, sport results in the occasional injury, but that applies to any participant whether or not he has haemophilia. The experience of other doctors who have had the opportunity to observe severely affected haemophilic children and adults during physical activity is that only one bleed is likely to occur for every 15 accidents!

More about sport will be found in Chapter 14.

At school	
He can	*He shouldn't*
Play in the playground	Box
Do physical education	Play rugby football
Swim	
Run	
Cycle	
Play football	
Go on school trips	
Do metalwork	
Do woodwork	

A typical school letter is shown on p. 322. Such a letter should set the scene for progress by emphasizing normality, and stressing the things the boy is able to do with his friends. It should start with some personal details about him. The diagnosis is then given and the fact that usual first aid measures apply is stressed. This is important because all teachers have knowledge of first aid and will be reassured. They don't have to learn anything new.

Some people still think that boys with haemophilia are likely to collapse in pools of blood and spoil the carpet, and the knowledge that most bleeds are internal is important. Because of this the teachers will need to know how to spot bleeds. The knowledge that the boy himself can tell when he is bleeding before there is anything to see will also be reassuring to them. Silly though it seems, some people also think that haemophilia can be caught. Teachers need to be told that it is not contagious and cannot affect other children, if only to reassure their parents.

A very important part of the letter is to explain who to contact should any problems arise, and how to do this. I append a card with the names of staff and telephone numbers to the introductory letter. In this way contact can be made directly with the centre if a parent is not available either at home or on a telephone pager. In the UK families with haemophilic children under the age of 16 years can benefit from a valuable free service called 'Armourpage', which is sponsored by the Armour Pharmaceutical Company in association with British Telecom and the Haemophilia Society.

The next section of the letter deals with activities and sports and again stresses normality. Normal development depends on continuing exploration of the world. Only the individual can eventually decide what is right for him in terms of work and leisure. Unnecessary restrictions hinder this progress. Children wanting to do the same things as their friends will eventually either reject restrictions directly, or hide the fact that they are disobeying their parents in order to avoid retribution. Far better to let them learn within sensible limits what is right for them and for their haemophilia. Finally, I ask to be kept up to date with the boy's progress. This is because I want to make sure that decisions which are unduly influenced by the diagnosis and which could affect his future are not taken at school.

In addition to the letter and the card, Haemophilia Society literature about schooling is helpful, as are some of the leaflets about haemophilia now available from the pharmaceutical companies. The text of an information leaflet we give to schools is shown in the box at the end of this chapter. Occasionally either the parents or the teachers ask for all this information to be followed up by a visit to the school by a member of the centre staff. This can be extremely helpful if there are special difficulties, for instance the presence of high-titre inhibitors.

Nowadays the great majority of boys with haemophilia can attend normal school and compete as equals with their peers. It is not very long ago that this was not the case, and in some countries it is still more usual for special schooling to be recommended for haemophilia. If this happens it is very important for the family to discuss the recommendation with centre staff and with representatives of the Education Authority before making a final decision.

The boy with haemophilia deserves the best education that is available. His future happiness and prosperity depend on schooling which has allowed him to thrive outside the diagnosis of haemophilia, which is after all only one of the many factors that influence decisions about an eventual career. The choice of that career is crucial to him. Although modern treatment allows him to do most jobs, whether they depend on strength of mind or strength of body, he will be competing with others without haemophilia. He has to prove to himself and to others than he is the best candidate for the job, with or without the bleeding disorder.

Main lesson of the day for this age group:
PLAN WELL AHEAD FOR HIS FUTURE!

Checklist

What teachers need to know:
His personal details
The diagnosis
The severity
First aid recommendations
Who to contact and how
What treatment is given
Will special provision be needed in school
Any effects on other children

Adolescence

It is a tribute to human resilience that most youngsters and their families survive adolescence. Being a teenager is difficult enough without the burden of haemophilia. Whilst parents need patience, a sense of humour, and a bottomless bank balance, their son needs both the security of home and the space to grow to the independence of adult life.

There is one word which sums up everything families need to know about adolescence and haemophilia. It is CONTROL. When someone is in control of their health, life is normal. When health is compromised and in control of them, life is abnormal. If someone with haemophilia allows his disorder to govern his life he is not in control. His haemophilia is, and his happiness is clouded. Learning to put life first and haemophilia second is the single most important feature of growing up.

To be in control requires discipline. It is useless to simply 'forget' haemophilia and not to, for instance, have a dose of factor VIII or IX before an event known to provoke a bleed. It is silly to imagine that a run of bleeds into a major joint will just 'go away' without therapy and without subsequent damage. It is foolhardy to go off on holiday without packing the equipment needed to treat a bleed, or knowing the location of the nearest haemophilia centre. In all these cases common-sense planning ensures that life can run smoothly without the disruption of untreated haemorrhage.

Most children will have learnt to control their haemophilia before puberty. They are in command, knowing exactly when to ask for treatment, judging the dose, mixing their concentrate, increasingly doing their own venepunctures, and clearing up afterwards. And then forgetting about it and getting on with the far more exciting events of life with their friends at home and school.

If this hasn't happened haemophilia can make adolescence very difficult for a while. At some stage there has to be a break. No one *wants* haemophilia, but it is there. Coming to terms with it, whilst at the same time wanting to slip the net of parental influence, can be very frightening and hurtful. The obvious response is anger, and anger used to be a frequent occurrence in the consulting room before comprehensive care became established and families learnt about letting children take responsibility for their haemophilia gradually as they grew up.

Careers

Choice of career is especially important in these days of high unemployment in many countries. Few jobs are closed to those with severe haemophilia. They include work involving particular hazard (for instance the armed forces and emergency services) or the risk of being isolated from special medical help should an emergency arise. With these exceptions and contrary to popular opinion, people with severe haemophilia are able to work manually, and some of them undertake the sort of work that would rapidly floor most doctors. Obviously, whatever career is chosen the man with haemophilia must be ready to compete as an equal to somebody without haemophilia, both when applying for jobs and when working.

There are obvious benefits in becoming as highly qualified as possible before leaving school or university; the choice is so much wider. All schools have easy access to advice about careers, but sometimes this is given rather late in the day. In the context of severe haemophilia it is important that the possibilities are explored sooner rather than later. Early planning helps to ensure that any disruption caused by untoward bleeding is catered for, and special help with difficult subjects is provided. Most youngsters need time to develop their own ideas about what they want to do with their lives. Early decisions can always be changed later in the light of new interests and work experience. When doubt about choice of career persists a consultation with an educational psychologist may help to point an indecisive teenager in the right direction.

Sexuality

The knowledge that haemophilia is inherited and that the abnormal gene can be passed on to children is bound to colour attitudes to sex and parenthood. Young men who have not yet come to terms with their haemophilia feel more 'different' than is usual in adolescence, and may find early courtship difficult because of this. Carrier girls are vulnerable because of the possibility of having haemophilic sons. Life is especially hard for them if there is a bad family history, or if Dad's haemophilia is complicated by severe arthritis, hepatitis, or HIV infection.

Two aspects of adolescence may need sensitive counselling. Firstly, carrier girls may have low factor VIII or IX levels and these may be linked to heavier periods than usual. When this happens it is very easy to put things right—one of the low-dose contraceptive pills, or regular courses of Cyklokapron, are all that is needed. Secondly, young men may have bleeds induced in the heat and passion of the moment. Again, treatment is straightforward, painful bruising being countered by a dose or two of the relevant factor concentrate. Whilst it is important to note that psoas bleeding is not always linked to sexual activity, a psoas bleed is fairly common at this age. This needs energetic treatment with replacement therapy for several days, together with bed rest and then physiotherapy. The psoas muscle runs from the side of the backbone and sweeps forwards round the pelvis to emerge in front of the thigh. If not stopped quickly bleeding into it compresses the major nerve to the quadriceps group of muscles leading to weakness, instability of the knee, and a loss of sensation to touch which can extend from groin to foot. Recovery from a poorly treated psoas bleed can take many months.

Sport and travel

Two aspects of normal adolescence need emphasizing. Like everything else in life, some youngsters like sport and excel in it and others do not. I know that it is tough on those who do not to have to tell them that exercise will be an essential part of their life with haemophilia, but there it is! It is *essential* for people with haemophilia to keep active in order to keep well. Physical activity promotes powerful muscles which protect the major joints. Flabby muscles cannot support joints properly and bleeding episodes are more frequent. Obesity adds to the problem by putting more strain on the unstable joints.

Travel during their teens allows most youngsters to make their first escape from parents and siblings. Holidays abroad, summer camps, and trips with the school all provide the necessary inducements. Haemophilia is not a bar to travel. Again, common-sense precautions ensure the trip is a happy one. A holiday letter detailing the medical history and the usual haemophilia treatment should be obtained from the centre, which should also provide a clearance note for Customs for any equipment, concentrate, or desmopressin (DDAVP) carried. A list of centres nearest to the destination, holiday insurance (read the small print!) and, in Europe, Form E111 are the other essentials. And don't forget that lads (and lasses) with the haemophilia gene are just as likely to fall foul of alcohol, drugs, and sexually transmitted diseases whilst footloose and fancy-free as anyone else.

Finally, having survived parents, brothers, sisters, schoolmasters, doctors, the dreaded nurse, AND adolescence, the average young man with haemophilia wants nothing more than to live his life his way. Great! But one thing to think about, please. Others less fortunate than yourself (especially those with haemophilia in developing countries) now need your enthusiasm, your energy, and your expertise in living with haemophilia. The various haemophilia societies and the World Federation of Hemophilia will welcome your active participation. So add them to your list . . . Greenpeace, Save the Whales, Antivivisection, . . . and the Haemophilia Society.

Main lesson for life:
KEEP IN CONTROL!

Background information
All you need to know about haemophilia in four pages

1. *Haemophilia*

   In this inherited disorder an ingredient called a factor, present in normal people and concerned with the clotting of the blood, is absent or partially deficient. The effects of this absence are apparent following injury when, because the normal clotting mechanism is impaired, the patient will continue to bleed. The duration of the bleeding will, of course, depend on the type of injury. Pin-pricks and small superficial wounds will bleed no more than in a normal person, because the tissues contract around the breach in the blood vessels. Larger cuts will bleed in exactly the same manner as in a normal person, but if left alone the bleeding will often (but not always) continue. Initial treatment is therefore easy, and usual first-aid measures apply. A dressing applied under pressure will temporarily stop the bleeding until treatment can be given.

   The most serious forms of bleeding in someone with haemophilia are those which involve the muscles, or joints. In a normal person blows and twists sometimes cause rupture of minute blood vessels resulting in bruising. The bruises are prevented from enlarging partly by the tension of the surrounding tissue and partly by the normal clotting mechanism. In the boy with haemophilia it is only when the pressure from the tissues is the same as the pressure exerted by the escaping blood that bleeding will stop. This is why untreated joints become swollen and painful. In the absence of proper treatment the bleeding takes a long time to resolve, the muscles round the joint become weak and, following repeated episodes, permanent disability can occur.

Background (cont.)

Treatment of bleeding in haemophilia is nowadays simple, and only requires an intravenous injection which replaces the missing factor in the blood. This injection may be given in the hospital or, when the family have been trained, at home. Boys with haemophilia know they are bleeding before there is anything to see. Treatment at this time usually stops the bleed immediately.

Occasionally an affected limb sometimes needs splinting, but it is rarely necessary to keep the splint on for more than a few days. There is no medical reason why a child should not attend school following all but the most serious injury, and it is helpful if permission for him to attend in a splint, or very rarely in a wheelchair, can be given by his teacher.

2. *General activities*

We feel that it is better that children with haemophilia require occasional treatment for bleeds, rather than that they should be banned from the normal activities of home and school. There is no reason why the child should not take part in activities like the school play or go on outings with the other children. Because of the risk of serious head injury we advise that activities like boxing and rugby football be avoided. We encourage other sports, including football, and advise that he should be allowed out in the playground with the other children, take part in activities like music and movement, and that he should be taught to swim. Swimming is a particularly good sport because it helps build up muscles and protect joints.

It is easy for those who know little about haemophilia to be frightened of the consequences of the normal high spirits of childhood. Severe bleeding is very rare indeed, and a balance must be achieved between protection and the aim of allowing the child to grow into a well-adjusted healthy adult.

3. *Indications for treatment*

- bleeding into a joint or a muscle causing swelling, pain, limitation of movement or sleeplessness;
- bleeding into a muscle;
- injury to the neck, mouth, tongue, or face;
- severe knocks to the head (see also below);
- heavy or persistent bleeding from any site;
- severe pain or swelling in any site;
- all open wounds requiring stitches;
- following any accident that may result in a bleed.

Nosebleeds are common and are liable to occur when infection is present. They can usually be controlled by applying pressure, with a finger, to the nose for 5 minutes. If bleeding is heavy or persistent further treatment is necessary.

If in doubt about whether treatment may be needed, seek advice from parents or haemophilia centre staff.

4. *Aspirin*

Aspirin is harmful to people with bleeding disorders, and should be avoided. It irritates the lining of the bowel and also acts directly on the blood clotting mechanism. In addition, aspirin has been implicated in a rare disorder of the liver in children, called Reye's Syndrome. As a consequence it is not recommended for use in children under the age of 12 years.

Paracetamol (Panadol) may be used instead of aspirin.

5. *Care of the teeth*

Good dental health is of the utmost importance to the child with a bleeding disorder. The extraction of a damaged tooth is often difficult, and always means special care.

Ideally the teeth should be brushed with toothpaste after every meal. If this is not practicable they should at the very least be brushed with toothpaste after breakfast and after supper. If a school dentist is available the child should be encouraged to see him regularly. An expert in dental hygiene is available at the haemophilia centre.

6. *Holidays*

Please tell the centre secretary if the child is going out of their hospital area for a school trip or holiday. You will be given the name of the haemophilia centre nearest to the holiday location, together with a letter of introduction (containing details of the child's disorder) for the local doctors.

7. *Identification*

The child must always carry a clear identification that he has a tendency to abnormal bleeding. This can be worn around the neck on a chain, or as a bracelet. Medic-Alert and Talisman identity discs and lockets are readily available. Medic-Alert identification carries a personal identity number (PIN) from which the Medic Alert foundation can access information about the individual. Talisman lockets contain relevant information on a paper strip.

8. *Immunization*

There is no reason why the child with haemophilia should not be immunized as usual provided simple precautions are taken. When vaccines are injected firm pressure must be applied to the injection site with a fingertip for a full 5 minutes after the injection.

9. *Infections*

Infections cause irritation of the tissues and may lead to bleeding. The family doctor should be seen as soon as an infection is apparent, or the haemophilia centre staff notified. Oral or intravenous antibiotics are suitable in haemophilia, but intramuscular injections should *not* be given because they may provoke deep bruising.

Background (cont.)

10. *Plasters or plastic splints*

    If a plaster fits too closely the following symptoms will occur:
    - tightness not relieved by elevating the limb;
    - marked swelling, blueness, or coldness of the fingers or toes;
    - pins and needles or inability to move the fingers or toes.

    In the case of back splints the pressure may relieved easily by loosening the retaining bandage. Plaster casts *must* be split open immediately, either by the boy's own doctor or at the hospital.

11. A NOTE ABOUT HEAD INJURY

    Head injury must always be treated seriously, and the child referred to the centre. Even after treatment has been given, immediate medical help is neccessary if any of the following occur within a two-week period:

    - unusual drowsiness, confusion, or failure to recognize surroundings
    - abnormal behaviour, or change of behaviour of a child
    - actual unconsciousness
    - persistent or severe headache or neck stiffness
    - repeated vomiting
    - weakness or loss of use of arm, leg, or face
    - blurred vision or seeing double
    - persistent discharge from nose or ears

    These notes are intended to give general information about haemophilia. Please let us know if they fail to answer any questions you have.

PART II

Medical Background

3

Bleeding and clotting

The suppleness of the healthy body provides its basic protection against abnormal blood loss. In movement tissues slide over each other. Joints are protected by strong tendons and ligaments. Their lubricated surfaces glide over each other. Healthy muscle and bone respond to the stresses and strains of normal exercise. Muscle tone and bulk protect deep structures, including major blood vessels. The vessels themselves, usually lying loose in these protective tissues, are pushed aside easily on all but the most sudden and severe impacts. When a vessel does break a series of events, known collectively as 'haemostasis' (literally static or still blood), begins.

Firstly, the injured vessel contracts and closes itself off from the circulation. Blood flow is shunted away from the site of the injury, allowing repair to take place in a low-pressure field. If the punctured vessel is small, this shutdown is all that is required to stop bleeding, at least for a while.

Secondly, a plug made of tiny particles builds up to stop the gap.

Thirdly, changes occur which convert liquid blood to solid clot. A permanent seal is formed.

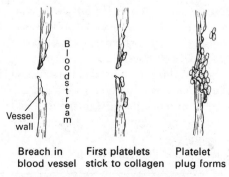

Breach in blood vessel | First platelets stick to collagen | Platelet plug forms

Fig. 3.1. Stages in clot formation.

The plug

The particles which make up the plug are called platelets. They are attracted by a substance called collagen which is found in the walls of blood vessels. Collagen is normally separated from flowing blood by a thin layer of cells which line the inside walls of vessels. When a break occurs in this lining blood comes into direct contact with collagen, and the first platelets to arrive on the scene stick to it. These platelets then release chemical substances which attract other platelets in the vicinity. The chemicals make the platelets stick to each other. Within seconds, a mass of sticky platelets is building up around the edges of the wound. Eventually enough platelets have collected to form a plug.

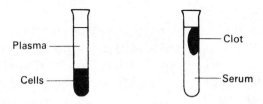

Fig. 3.2. The difference between plasma and serum. In the first tube the blood is fluid and the cells have settled out. In the second tube the blood has clotted.

Background
Blood

Although to the naked eye freshly drawn blood looks completely fluid, it contains millions of cells. The cells float in a liquid called plasma. Plasma is like a very rich soup; it contains many important proteins, including those necessary for normal clotting and for the defence of the body. When blood clots some of these proteins are used up and the resulting fluid, thinner than plasma, is called serum.

The cells of the blood are of various types, each with a special function. The most numerous are the red cells, which contain a substance called haemoglobin. Haemoglobin has the remarkable property of being able to combine with oxygen, to hold on to it until cells requiring oxygen are reached, and then to give it up in exchange for waste carbon dioxide which passes into the blood for transport to the lungs. When combined with oxygen, haemoglobin becomes bright red; hence 'red' cells and 'red' blood. There are about six million red cells in every cubic millimetre of a man's blood—a cubic millimetre is about the size of an 'o' on this page.

If either the number of red cells falls, or the haemoglobin in them is reduced, the patient becomes pale and is said to be 'anaemic'. The most usual causes of anaemia are either blood loss or iron deficiency, because iron is necessary for the body to make haemoglobin. Not all pale people are anaemic—they may just have thick skins which hide their capillaries more effectively.

Red cells are made in the marrow of the bones. Children need to make many red cells to keep up with their growth and all their long bones are full of red-marrow factories. Adults need only to keep up with the natural death-rate of red cells within their circulations, each red cell living for about 120 days. When red cells die they release their haemoglobin which is broken down into iron, which is used again by the body, and the pigment from which the liver produces bile. In adults red cells are made mainly in flat bones which include the breastbone and pelvis.

On the surface of the red cells are specialized areas called antigen sites. By identifying red cell antigens in the laboratory the red cells from different people can be typed into a number of different categories. Two of the most important of these categories are called the ABO system and the Rhesus system. Human blood can be divided into groups A, B, AB, or O and into Rhesus positive or Rhesus negative. This is done with all blood donors and anyone needing a transfusion, as well as in all the best detective stories when the blood from the scene of the crime is compared with the blood from the suspect's cut finger.

Blood grouping is of vital importance in medicine because the body recognizes substances foreign to itself and destroys them by forming antibodies. Thus if a person with blood group A is transfused with group A blood all is well; the bloods match. If the same group A person

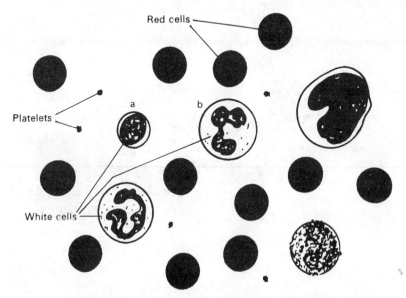

Fig. 3.3. Diagram of human blood cells (magnified about 1000 times). The cell marked (a) is a lymphocyte; that marked (b) is a granulocyte.

Background (cont.)

is transfused with group B blood his body recognizes the invasion of a foreign antigen (B) and sets about destroying it with an antibody, with unpleasant consequences. Plasma and its derivatives do not contain these antigens but, because a few red cells may be left in fresh frozen plasma during the manufacturing process, some doctors like to give their patients group-specific plasma or its products rather than plasma of any group.

There are two other sorts of cell in the blood. Platelets help control bleeding, and white cells help protect the body against harm from infections and other invaders. Platelets, or thrombocytes, are the smallest cells in the blood. There are about 200 000 of them in a millilitre of blood; they are made in the bone marrow. Like red cells, platelets can be affected by antibodies. If platelet antibodies occur, their attachment to the cells is recognized by the spleen, an organ tucked under the ribs on the left side of the body. One of the functions of the spleen is to remove foreign material, and it soon recognizes platelets damaged by antibodies and eliminates them. The result is a fall in the number of circulating platelets, a condition known as thrombocytopenia. Because we do not always know why antibodies have appeared, the word *idiopathic* (literally of 'private' or unknown cause) is used to describe the condition that results.

A lack of platelets results in a failure of blood vessels, especially when under tension, to retain red cells. They escape into the tissues, appearing as red dots in the skin, a condition called purpura.

Information about white cells will be found in Chapter 12.

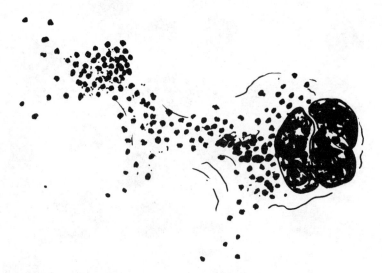

Fig. 3.4. Diagram of the birth of blood platelets from the mother cell in the bone marrow (magnified about 2500 times).

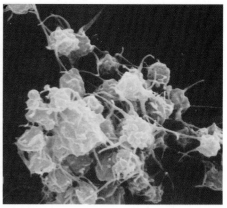

1. In response to local injury two plate-lets reach out their pseudopodia to stick together.

2. More platelets join in.

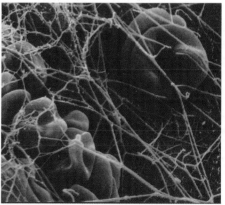

3. The platelet plug builds up to stem bleeding. In the background are red blood cells.

4. After a while strands of fibrin appear and strengthen the repair.

Diagrams of what happens inside a blood vessel when its wall is damaged.

Photomicrographs 1–4 by Dr Lennart Nilsson of the Karolinska Institute, Stockholm.

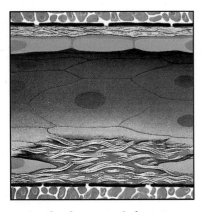

5. At the bottom of the picture collagen fibres have been exposed by injury; at the top are the normal living cells.

6. A few platelets stick to the collagen.

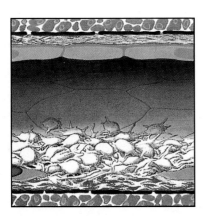

7. Platelets stick to each other building up a plug.

8. Clotting results in the appearance of fibrin strands which bind everything together and stop the bleeding permanently.

Diagrams of what happens inside a blood vessel when its wall is damaged. (contd.)

Colour plates 5−8 courtesy of Boehringer Ingelheim and Alpha Therapeutic.

The contraction of blood vessels and the formation of the platelet plug are the first lines of defence once bleeding starts. This stage is illustrated very nicely, together with the formation of a clot, in the colour plates and diagrams.

Some disorders of blood vessels and platelets

The vessels and platelets may be altered in a variety of ways with consequent bleeding. This bleeding is different from that experienced in haemophilia A and B, in which the first lines of defence are normal. It is immediate and often profuse and continues for longer than normal. The bleeding time, which is the time taken for bleeding to stop after a pin-prick in an ear lobe or small cuts in a forearm are made, is typically prolonged. People with severe vessel and platelet disorders tend to bleed from the nose and bowel, and women have heavy periods. Joint and muscle bleeds are rare. One of the common manifestations of these disorders is the appearance in the skin of crops of tiny bleeds, the result looking like a reddish-blue rash. The medical name for this is purpura. A purpuric rash fails to disappear with pressure, unlike 'nettle rash' or the rashes associated with the infectious fevers like measles. Sometimes many purpuric spots run together to form a bruise.

A lack of platelets, or a failure in their function, produces purpura because tiny gaps in blood vessel walls are not sealed with effective platelet plugs. These gaps may result from very slight injury, such as the pressure of a tight sweater, or from a sudden increase in the pressure of blood inside a vessel, as occurs for instance when coughing or when

Fig. 3.5. Purpura on a child's arm.

straining on the toilet. They are also seen in conditions associated with fragile blood vessels, the commonest of which is old age, when the skin and the tissues supporting the vessels have become very thin and frail. Some people are born with blood vessels a little more fragile than usual. They experience a sensation very like that felt when an insect bites as a small vessel pops and they start to develop a bruise. In medieval times their bruises were thought to be caused by witches or the Devil himself, and they were called 'Devil's pinches'.

Whilst fragility is a natural consequence of growing old, it is sometimes associated with illness in younger people. Of historical interest is the disease scurvy. Scurvy used to afflict whole armies, and was particularly common in men who sailed the seas in the great voyages of exploration. The cause and cure were discovered by Dr James Lind of the Royal Navy in 1747, and confirmed by Captain Cook on his journey to New Zealand 25 years later. Scurvy causes profound changes in the tissues of the body, including those of the blood vessels, which weaken and rupture. It is caused by a lack of vitamin C (ascorbic acid). Lind observed that sailors given citrus fruits, which we now know are rich in vitamin C, in addition to the then usual diet of saltmeat and biscuits, recovered from their affliction. The Royal Navy effectively doubled its fighting strength by including fresh fruit juice in its rations, a procedure that led to the nickname 'limeys' for British sailors.

In addition to its effect on blood vessels, scurvy also alters the mechanism by which platelets stick to each other. With increasing knowledge of platelet function doctors are recognizing more conditions in which this stickiness is reduced. Some, like thrombasthenia (Glanzmanns disease), are inherited, and others are secondary to other diseases or drugs. The most important of the drug causes is aspirin, and this is one of the reasons why aspirin is so strongly condemned in the treatment of anyone with a bleeding disorder.

Hereditary haemorrhagic telangiectasia

This complicated name refers to a disorder in which the great majority of vessels, and all the platelets, are normal, but now and again little red knots of tortuous small vessels appear in the tissues. The knots are about the size of pinheads and may sometimes be seen on the lips and fingers (Fig. 3.6). Here they cause no trouble, but when they appear in the linings of internal organs they may be the site of bleeding. The tendency to develop these knots, or telangiectases, is inherited, but they are rarely a worry until after puberty.

Unfortunately in a susceptible person they continue to form, so that the removal of troublesome ones by surgery or laser treatment does not

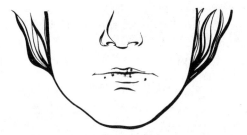

Fig. 3.6. The lesions of hereditary haemorrhagic telangiectasia.

provide a cure. However, the condition is not often severe; indeed, the lesions are remarkably common in the population. The complication usually encountered is frequent nosebleeds, and the occasional unlucky person requires blood transfusion for the consequent anaemia.

Ehlers Danlos syndrome

In some forms of this inherited condition blood vessels are unusually friable and may give way easily. The most common effect of this fragility is easy bruising. People with Ehlers Danlos syndrome are often double-jointed and the contortionists in circuses are obvious examples of the disorder. They are able to perform their tricks because one of the constituents of many body structures, collagen, has been constructed in an abnormal way. There are many varieties of this syndrome and they can only be sorted out by taking small samples (biopsies) of the skin and examining their collagen content. Differentiation is important because different forms of Ehlers Danlos syndrome are inherited in different ways.

Clotting and clotting factors

At first sight the clotting reactions, which form the last line of defence against loss of blood, look as simple as the housewife's recipe for a cake. Just as a cake is made from ingredients in the kitchen, a clot may be made from ingredients in the laboratory. Instead of the flour, butter, and eggs needed for the cake the clot requires ingredients called **factors**.

The main clotting factors

There are 12 main clotting factors. By international convention they are numbered in the Roman manner (I, II, III, IV, etc.). The number VI was originally given to a substance thought to be derived from factor V; subsequent investigation disproved this idea and VI no longer appears in the list of factors.

The factors, together with some of their alternative names, are:

   I fibrinogen
  II prothrombin
 III tissue extract
 IV calcium
  V
 VII
VIII anti-haemophilic factor (AHF) anti-haemophilic globulin (AHG)
 IX Christmas factor
  X
 XI plasma thromboplastin antecedent (PTA)
XII Hageman factor
XIII fibrin stablizing factor

The names Christmas and Hageman were the names of the first people to be discovered with a deficiency of factors IX and XII, respectively.

As our knowledge grows, other factors or variants of known factors come to light. Already it is possible to identify the exact site of abnormal change in some of the factors and to link the changes with the effects in particular patients. A specific change in molecular structure is often identified with the place of its discovery. Hence we have 'fibrinogen Amsterdam' and 'prothrombin Madrid'.

What are the factors?

With the exception of III and IV, the factors are proteins manufactured in the body. Most of them are made in the liver.

Factor III is a simple term for a complex collection of substances present in the fluid that surrounds cells. It is released from here, and from the mashed cells themselves, on injury. Factor IV is the term for particles of calcium that carry an electrical charge.

The body builds the protein factors from units digested from the food. Four of these factors (II, VII, IX, and X) require vitamin K for their manufacture by the liver. Vitamin K is present in a large variety of animal and plant foodstuffs; no one on a normal diet is short of it. A lack may occur in premature or some breastfed babies and people with liver or digestive diseases, and in these cases the vitamin will be prescribed by the doctor to correct the deficiency and prevent bleeding.

Two further vitamin K dependent proteins called Protein C and Protein S act as anticoagulants in the regulation of clotting, and inherited deficiencies may result in thromboses.

The life of the factors

The normal body is continually producing clotting factors in order to keep up with natural loss. In time of need this production is stepped up for a while to cover real or anticipated increases in the use of factors. This is seen following injury, and during pregnancy when good clotting is required to protect the mother at the birth of her baby.

Once a factor has been released into the bloodstream its clotting activity gradually falls away. The rate of the fall, which is caused by natural destruction or decay, varies from factor to factor. This rate may be measured in the laboratory, and for convenience the concept of 'half-life' is used when describing it. The half-life is the time taken for half the clotting activity, as present initially after transfusion, to disappear from the blood of a patient who is not bleeding at the time.

For factor VIII the half-life is 12 hours. This means that 12 hours after the recording of, say 100 per cent activity in a blood sample, only 50 per cent will remain. Twenty-four (12 + 12) hours after the initial measurement there will be 25 per cent clotting activity; 36(12 + 12 + 12) hours after, 12.5 per cent activity; and so on. When plotted graphically the results are as shown in Fig. 3.7.

By knowing the expected half-life of a factor's clotting activity the

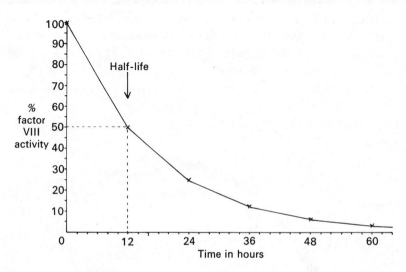

Fig. 3.7. The decay of factor VIII activity at body temperature (decay is stopped if plasma is deep frozen).

doctor can calculate how often someone deficient in the factor will need treatment to prevent bleeding after injury or surgery. The doctor will also be able to calculate the amount of blood product likely to be needed before embarking on an operation.

Thus it is important to understand that the clotting factors are not permanent, like eyes, or ears, or hands. They are constantly being recycled by the body, each 'piece' produced surviving for only a few hours before its activity dies away. This is why haemophilia and the other inherited clotting disorders cannot be cured permanently by changing the blood or transfusing the missing factor on one occasion only.

How do the factors work?

If freshly drawn, normal blood plasma is examined under a microscope, there is little to see at first, but before long thin strands of material begin to form in the fluid. Within a few minutes enough of the strands have formed to give an appearance like a nylon mesh. At this stage the liquid plasma has become solid; it has clotted. The time taken between the withdrawal of blood and the appearance of a clot is called the clotting time.

The strands are called fibrin. In the terms of the cooking analogy, fibrin represents the baked cake. Like the cake, fibrin is the end product of a series of reactions amongst the ingredients.

In man there are two starting-points for these reactions and both are triggered by injury when a blood vessel is broken. The first follows the release of factor III (tissue extract) and its mixing with the escaping blood. In the blood is factor VII. Before the injury, factor VII, together with the other circulating factors, has been dormant. Acting like an alarm clock, factor III now wakes, or 'activates' factor VII, and a chain reaction culminating in the formation of fibrin begins (Fig. 3.8). Some descriptions of the clotting process liken this chain

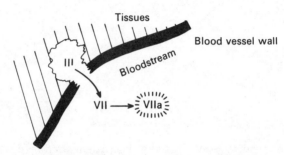

Fig. 3.8. Diagram of the first part of the extrinsic clotting system.

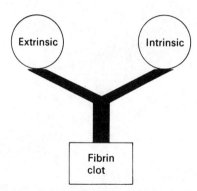

Fig. 3.9. The extrinsic pathway (starting outside the bloodstream) acts with the intrinsic pathway (starting inside the bloodstream) to produce a fibrin clot.

reaction to a sequence of falling dominoes. If one domino is missing, the sequence fails.

This pathway to clotting is called 'extrinsic' because it starts, with factor III, outside the bloodstream.

The second start point for clotting (Fig. 3.9) involves a direct response in the bloodstream itself, and the resulting pathway is called 'intrinsic'. Here stocks of ingredients are continuously circulating throughout the body in the blood, so that a proportion of the correct materials is always near the site of any injury.

The intrinsic pathway is activated when blood comes into contact with collagen, the substance laid bare when the inside lining of a blood vessel is damaged. The first factors to be affected are XII and XI. Contact with collagen activates them and another series of reactions begins. The reactions proceed like a stepped waterfall until fibrin is reached.

The first steps of the waterfall (Fig. 3.10) include the haemophilia factors VIII and IX. In the diagrams 'a' means 'activated'.

The later steps of the waterfall (Fig. 3.11) are common to both the extrinsic and intrinsic systems; they form a final common pathway which ends in fibrin. It is during the last reaction, in which factor I (fibrinogen) is converted into fibrin, that the strands become visible under the microscope.

Factor IV (calcium) is required at several steps of the waterfall, and is not shown in the diagram. Factor XIII stabilizes and strengthens the fibrin to prevent it breaking down too soon.

In a healthy person the four systems that stop bleeding—contraction of the blood vessels, formation of the platelet plug, the extrinsic pathway, and the intrinsic pathway—act together with many cross-links between them. The pressure of blood required in a person's vessels in order to achieve a satisfactory circulation is too great for either vessel

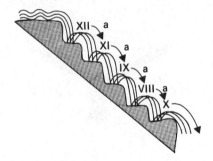

Fig. 3.10. Activation of clotting factor is like a waterfall.

contraction or the platelet plug to be effective in stopping bleeding from all but the smallest injuries, although they stem the flow until clotting occurs. The extrinsic pathway provides some fibrin, but tissue extract from the site is used up too rapidly to form a permanent seal to hold the pressure, and it is here that the intrinsic system takes over. It does this by adding fibrin from its pool of blood-borne ingredients which are continuously replaced by the body.

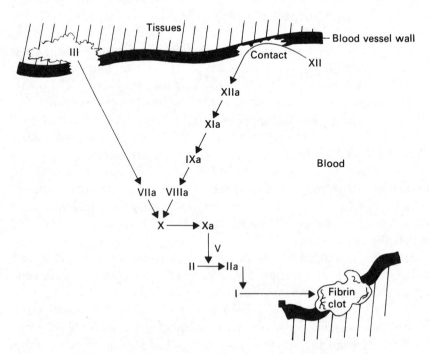

Fig. 3.11. Diagram of the activation of the clotting factors to form a fibrin clot (calcium—factor IV—and platelets act at several stages).

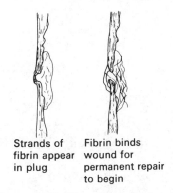

Strands of   Fibrin binds
fibrin appear wound for
in plug   permanent repair
               to begin

Fig. 3.12. End result of clotting.

The end result is a masterpiece of mechanical engineering, strong enough to resist the blood pressure, yet pliable enough to withstand movement. Its strength is provided by the fibrin strands which interlace the platelet plug in the same way that iron rods reinforce the mortar in ferroconcrete (Fig. 3.12).

Once bleeding has stopped, the firmly adhering clot slowly contracts, drawing the edges of the wound together. Body cells then gradually replace the clot and tough scar tissue effects a permanent repair.

Research into how and why blood clots in the way it does is of fundamental importance to the health of far more people than those born with a clotting deficiency. Whilst only one in 10 000 people has haemophilia A, about half of all deaths in the general population involve abnormal clotting. Thus, knowledge of what happens in the haemophilias and von Willebrand disease is of great interest in the study of coronary artery disease. Conversely, research into the causes of thrombosis helps our understanding of factor deficiencies.

We know that the clotting cascade or waterfall is a very simplified concept of what actually happens when blood begins to congeal. Rather than acting individually, factors have to work very closely together in teams in order for clotting to proceed normally. Blood platelets are vital members of the different teams, providing specially prepared surfaces on which the reactions take place.

How factors are activated becomes clearer as we learn more about their structures. It seems that the body produces the factors complete with safety caps which must be removed by chemical action in order to activate the underlying structures. For instance, factor IXa, with the aid of its team, bites pieces off factor X in order to reveal its active sites and to allow it, in turn, to begin work.

The actions of factor IX and the other factors present in intermediate

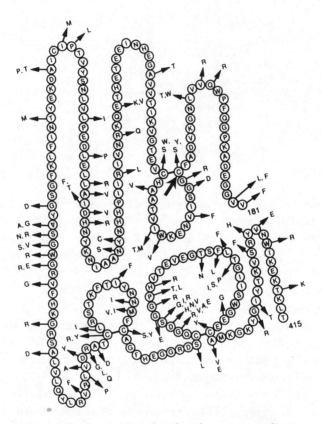

Fig. 3.13. Part of the factor IX molecule. The arrows indicate mutations from the normal sequence. It is these mutations which have caused haemophilia B in the individuals concerned. The first person to be diagnosed as having haemophilia B was called Christmas (hence 'Christmas disease'). His mutation is marked in the diagram which is reproduced with the kind permission of Dr Harold Roberts of Chapel Hill, North Carolina, USA.

purity concentrates, especially factors VII and X, are of great importance to people with factor VIII inhibitors. Links between the extrinsic and the intrinsic clotting pathways allow for the bypassing of factor VIII, and the direct activation of the final common pathway to fibrin. These links may also be responsible for the inappropriate clotting sometimes induced in people with haemophilia B who are receiving treatment with factor IX concentrate. The new high-purity concentrates which contain only factor IX have been designed to overcome this problem.

The unravelling of these processes and the structures involved provides a great challenge to researchers because once the blueprints of a clotting factor are known it becomes possible to build a synthetic model.

Fig. 3.13 shows something of the complexity of one of these biological assemblies. It is a diagram of factor IX, made up of a series of particles called amino acids. The amino acids are arranged in an exact sequence for the assembly to work in the body. Haemophilia B is the result of an incorrect or incomplete sequence. The long, winding chain of amino acids shown in the diagram is folded up in a very precise way and held together at specific points by electrical and chemical links. Although factor VIII is also made of chains of amino acids which have been folded and compressed in a similar way, it is a much larger and more complex structure than factor IX (Fig. 3.14).

In order to explain the differences between von Willebrand disease and haemophilia A we know that factor VIII is made in at least two pieces (p. 85). One of these, which is affected in von Willebrand disease, has been identified in the lining of blood vessel walls and in platelets and their parent cells. The other, smaller piece (VIII: C) is affected in haemophilia A. To be effective both pieces of factor VIII must link up in the bloodstream, and in the knowledge of how powerful the effect of just a little active factor VIII is in clotting, one might expect a fairly robust molecule. Instead its activity is so delicate that it disappears rapidly when fresh blood is handled in a blood bank or laboratory.

One result of this frailty was that until recently we did not know where in the body VIII:C is made. This butterfly of clotting has now been caught. Factor VIII:C is made in cells called hepatocytes in the liver.

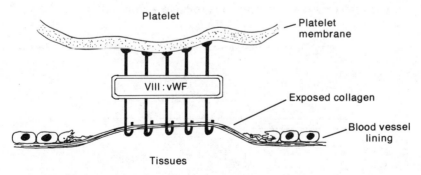

Fig. 3.14. Factor VIII is a very large molecule because one of its tasks is to stick platelets to damaged blood vessel walls. The glue which does this is in the von Willebrand factor part of the assembly.

Safety precautions

Anyone reading about the complexity of blood clotting (and the above description is very simplified!) may naturally ask: 'Why so many factors?' After all, it would be far simpler to activate factor I (fibrinogen) directly.

The answer lies in safety. If instantaneous clotting occurred on injury, a person would be in great danger. The stresses and strains of life mean that the body is constantly having to repair itself. If the smallest injury resulted in a powerful explosion of fibrin, major vessels would soon be blocked and vital organs shut off. The long sequence of factors provides the possibility of cut-off points at several stages. Clotting is full of cut-offs and fail-safe mechanisms, and it is because of them that a clot usually remains localized to the site of injury and does not extend into the circulation.

The most important safety mechanism involved is that of fibrinolysis. This mechanism is responsible for the removal of unwanted fibrin. Within each clot are fibrinolytic substances which are activated in a way similar to the activation of the clotting factors. They destroy fibrin by breaking it up or 'lysing' it, and the resulting fragments are removed safely in the bloodstream.

Every clot, therefore, contains the seeds of its own destruction, and a delicate balance is maintained so that both bleeding and excess clotting are prevented (Fig. 3.15).

Fibrinolysis may be blocked with certain drugs (p. 137), and this blockade is very useful in the treatment of some haemophilic bleeding, especially that associated with dental extractions. The drugs prevent the early breakdown of clots formed with the aid of factor replacement, and considerably shorten the time otherwise required for treatment.

There is one further point of interest before leaving the list of factors. It would be very wasteful for the body to have to maintain large stocks of

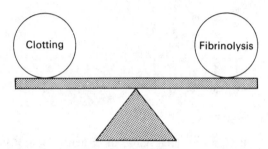

Fig. 3.15. The delicate balance between clotting and fibrinolysis (breakdown of fibrin).

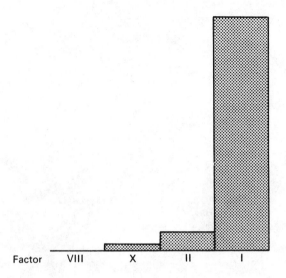

Fig. 3.16. Diagram of the different proportions of four of the clotting factors in the blood.

each ingredient in the circulation. Only tiny amounts of the early factors are required to create a mass of fibrin at the site of injury. Each of the factors activated is present in the bloodstream in a larger amount than the one before. By the time factor I (fibrinogen) is reached, the chain reaction has built up, or amplified itself, to have the required effect. This is shown in Figs 3.16 and 3.17, and is vital to understanding the modern management of haemophilia. If the chain reaction cannot proceed because one of the links is deficient, enough fibrin cannot be laid down quickly enough to stop the bleeding. Fast replacement of the link restores the chain immediately and normal clotting occurs.

Disorders of clotting

The baker's cake will be spoilt if she has forgotten to add one of the ingredients required by the recipe. Similarly, a blood clot will be defective if one of the clotting factors is absent or deficient, and the clotting time is usually prolonged. Thankfully the majority of people born with a bleeding disorder have an isolated factor defect, and people in whom more than one factor is deficient are very rare.

The commonest of the hereditary bleeding disorders due to a single factor deficiency is haemophilia A in which there is a lack of active factor VIII. Haemophilia A accounts for about 85 per cent of these disorders and haemophilia B (Christmas disease), which is caused by a deficiency in factor IX, for about 15 per cent. Isolated defects of factors

Fig. 3.17. Progressive build-up in a chain reaction of factors results in a virtual explosion of fibrin just where it is needed—at the site of injury.

I, II, V, VII, X, XI, XII, and XIII are occasionally found. Their numbers tend to vary according to race and locality. For instance, people with factor XI deficiency are often Ashkenazi Jews.

The incidence of haemophilia A also seems to vary from country to country, but this simply reflects how carefully the milder forms of the disorder are looked for in the population. Haemophilia affects about one person in every 10 000. Some 6000 people with haemophilia are known in the United Kingdom, and there are probably 25 000 in the United States.

The disorder known as von Willebrand disease, in which there is both a deficiency in factor VIII and in platelet function, with consequent prolongation of the clotting and bleeding times, is very common in some communities because of its inheritance (p. 78). In these communities the disorder has been passed down several branches of a family, the members of which may have grown up, settled in the same locality, and intermarried.

Severity, activity, and protein

In recent years it has become apparent that there are different forms of haemophilia A and B, and von Willebrand disease, and different forms of the other factor deficiencies as well. When referring to the 'severity' of haemophilia the doctor is correlating the clotting activity of factor VIII or IX and the manifestations of the disorder in his patients. In general, those with no (0 per cent) factor VIII or IX activity bleed spontaneously and are especially liable to deep muscle and joint bleeds; they are said to be severely affected. Those with between 1 and 5 per cent activity are more moderately affected, bleeds usually being related to injury, while those with more than 5 per cent activity are said to be 'mild'.

This relationship between the activity level and the degree of injury required to cause bleeding is useful when prescribing treatment for patients. This is shown in Fig. 3.18, which also shows that the normal activity levels are between 50 and 200 per cent, the average level in the population as a whole being 100 per cent.

A more modern terminology for the level of the clotting factors is now used in the medical literature. It relates the amount of the factor directly to the volume of blood containing it. As 100 per cent is one unit in the new system, translation is easy (Table 3.1).

That the relationship between activity level and clinical severity is

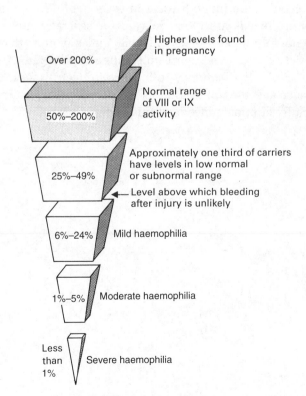

Fig. 3.18. The range of factor VIII or IX activity.

Table 3.1 Systems of measuring clotting
activity

Old system (percentage)	New system (units in a millilitre of blood; U/ml)
200	2.00
100	1.00
50	0.5
25	0.25
10	0.10
5	0.05
1	0.01

not so straightforward is recognized by those with experience in looking after haemophilia. Some boys with no apparent factor activity never suffer severe bleeds, whilst others, who by their factor levels should be 'mild', have continual problems.

The character of the bleeding in the clotting disorders is explained in greater detail in Chapter 5. Because the first lines of defence against bleeding are intact (except in severe von Willebrand disease), haemophilic bleeds are often delayed for several hours after injury. They do not follow pin-pricks and small cuts any more than in normal people, but after greater injuries bleeding is prolonged. An important characteristic is the tendency to bleed into deep tissues, particularly the muscles and the joints. But, as described later, modern treatment can virtually eliminate bleeds of this sort.

4

The causes of bleeding disorders

The human body grows from a single cell, the fertilized ovum. At conception this cell is formed by the union of the father's sperm and the mother's egg. Within the cell lies the nucleus which governs all the functions of the cell. Packed within this first nucleus, a structure so small that it can only be seen through a microscope, is all the information needed for the control of the growth and development of the baby. This information has come from the child's father and mother and, through them, from previous generations of the family. The nature of this information is explained later in this chapter. When a single piece of the information is faulty, a bleeding disorder may result.

The growing baby

As the cell and its daughter cells divide, and divide again billions of times, the cells begin to differ from each other. The result is a body made of cells of many sizes, shapes, and functions.

By the end of the fourth week from conception the heart is beating,

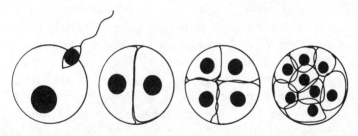

Fig. 4.1. Fertilization of ovum by sperm and early division of first cell.

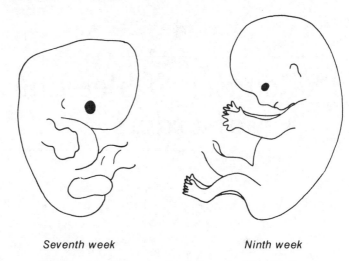

Seventh week   *Ninth week*

Fig. 4.2. Stages in development of the fetus.

the eyes and ears have started to form, and the legs and arms look like little buds. In the fifth week, at the time the mother may start to experience morning sickness, the face and brain are forming. By the twelfth week the baby's sex is apparent, nails have grown and the kidneys have begun to function. At 16 weeks the baby, or fetus, is fully formed. Hair has started to grow, and the muscles can contract. The mother feels the first fluttering movements of life within her.

The remainder of pregnancy is necessary for growth. At 16 weeks the baby is only about 15 centimetres (6 inches) long, as compared with the full-term baby who at birth measures 50 centimetres (20 inches). In the first 16 weeks of pregnancy harmful substances can affect the development of the future child. Change is so rapid that a single dose of a harmful drug may result in a devastating block in the development of a particular organ. This happened with thalidomide and can also follow certain of the infectious fevers. After 16 weeks, ingestion of harmful substances, including alcohol, affect growth. Cigarette smoking is especially harmful. The babies of mothers who smoke are smaller than the babies of mothers who do not. We now know that this effect happens even when the mother gives up cigarettes in early pregnancy. The message is clear. Women intending to become pregnant should stop smoking before conception.

Although there is no evidence that drugs taken during pregnancy result in a bleeding disorder in the child, it is sensible to avoid all drugs in early pregnancy, including those bought from the chemist without prescription. The child with a bleeding disorder can well do without an additional handicap.

How cells work

We are all made of cells. Cells are like miniature factories (Fig. 4.3), continually producing all the substances necessary for life. The management of the factory is in the nucleus. From here instructions are sent to the shop floor, or cytoplasm, where materials derived from food are assembled. The energy to drive the whole process is created in the cell factory by the breakdown of chemicals.

Within the managerial nucleus is a library of information needed to run the factory. This information is stored on 'shelves' called chromosomes. Each piece of information on the chromosomes is called a gene.

Using special staining techniques, the chromosomes can be seen under a microscope. In each of the body cells there are 46 chromosomes. When a cell divides the chromosomes divide too, so that each of the new cells contains just the right number. The time of division is the best time to see the chromosomes; the pictures show them in the act of division. Before they separate each pair of daughter chromosomes is held together at one point, called the centromere (Figure 4.4).

If a photograph is taken of the view through the microscope and the chromosomes are cut out of the resulting print, the cut-outs can be arranged according to their lengths and the positions of the centromeres.

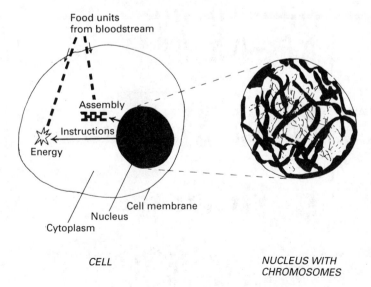

CELL

NUCLEUS WITH
CHROMOSOMES

Fig. 4.3. A cell is like a miniature factory with the management found in the nucleus.

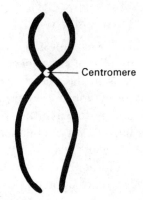

Fig. 4.4. Human chromosome.

This has been done in Figs 4.5 and 4.6. One shows the chromosomes from a male, and the other the chromosomes from a female.

The patterns for 44 of the chromosomes are the same for both sexes. They are arranged neatly in pairs. The only difference between the pictures is found in the last pair. The chromosomes from the female look identical. Those from the male look different. One is like those of the last pair in the female. The other is very small.

The first 44 chromosomes are called autosomes and the remaining two are called sex chromosomes. The bigger sex chromosome is named

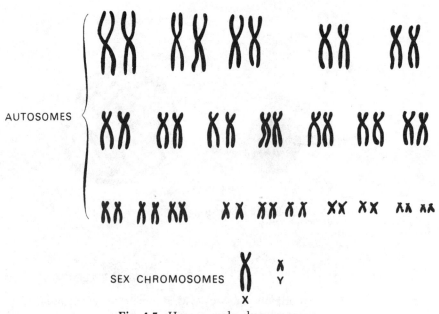

Fig. 4.5. Human male chromosomes.

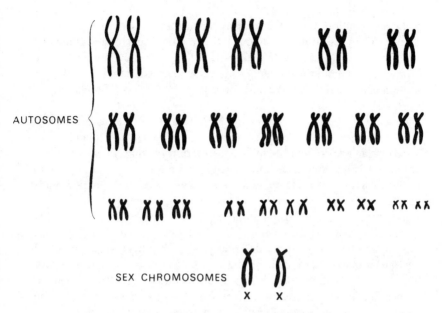

AUTOSOMES

SEX CHROMOSOMES
X   X

Fig. 4.6. Human female chromosomes.

X, the smaller one Y. Females have two X chromosomes (XX) and males one X and one Y (XY).

The neatness of the pairing of the chromosomes is no accident. The fertilized ovum results from the union of a sperm from the father and an egg (ovum) from the mother. Both sperms and eggs are the results of cell division, but in the reproductive organs this division is different from that taking place in other body cells.

It is easy to see why this should be. When a body cell divides into two, each of the new cells contains 46 chromosomes. If a sperm contained 46 chromosomes and an egg 46 chromosomes, the resultant first cell of the baby would contain 46 + 46 = 92 chromosomes. A very odd baby indeed!

Background information
Genes

The genes form part of the DNA (deoxyribonucleic acid) molecule, which exists in the form of a double helix, which is in turn coiled up on itself to form a chromosome. It is intriguing that the genes, which control all living things, only account for around 5 per cent of the DNA. The remaining material, which separates the genes, appears to have no function. Similar 'gaps' called introns occur in the genes themselves, where they separate the coding sequences or exons from each other. Genes work by using those coding sequences to make proteins. The sequences are constructed

Background (cont.)

from combinations of only four chemical building blocks which between them provide the 'alphabet' or 'code' of life. They are called adenine (A), thymine (T), cytosine (C), and guanine (G). In the gene, these blocks are arranged in pairs, joined together by weak chemical bonds (Fig. 4.7). A fundamental rule is that A bonds only to T, (A–T) and C bonds only to G, (C–G).

It is the order of the blocks in the length of DNA that determines what is made. Three blocks in a row (a codon) are the blueprint for a particular amino acid, or for an instruction on how to put different amino acids together. Once assembled, the amino acids form a protein. Clotting factors are proteins, and are simply a sequence of amino acids constructed from a blueprint provided by the DNA gene.

In order to produce finished proteins, the ladder-like DNA spiral splits at the sites of its weak chemical bonds, like a zip fastener (Fig. 4.8). The order of these split bonds is copied exactly by a single-stranded messenger called RNA. This strand is then fed into a structure within the cell that translates the message it carries into amino acid sequences. It does this by assembling the corresponding pieces to each codon of the message. Once the length of protein is complete it folds up into a precise shape, like Japanese origami. It can now take its place in the regulation of the cell's function.

The sum of the enormous amount of information stored in our genes is called the genome. In humans, the genome contains about three billion pieces of information, and scientists are presently mapping them all so that the entire sequence of the instructions for human life will be available in the form of a genetic library. Gene mapping is done using probes which are made to identify specific sequences of instructions on each chromosome.

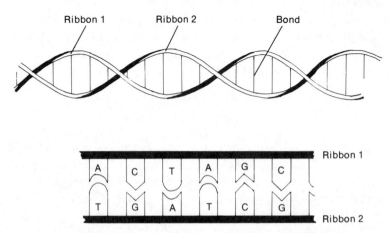

Fig. 4.7. The two ribbons of the double helix of DNA held together by chemical bonds between adenine (A), thymine (T), cytosine (C), and guanine (G).

Already our knowledge of the structure of individual genes and how they work has led to a revolution in medicine. By using the process called recombinant DNA technology, in which foreign DNA is combined with the native DNA of a living cell, we are now able to instruct cells to build biologically active products to order. The list of products already in clinical use is impressive. The bacterium *Escherichia coli* is used to make insulin for the treatment of sugar diabetes, growth hormone for the treatment of some forms of dwarfism, and interferon-alfa, a drug used in viral and malignant disease. Yeast cells are used to make the synthetic form of the hepatitis B vaccine, Engerix B, and hamster cells have been harnessed to make TPA (tissue plasminogen activator, Actilyse) which combats coronary artery thrombosis, and erythropoietin which stimulates red cell production.

In terms of haemophilia, the knowledge has not only resulted in the production of recombinant factor VIII (p. 130) but has also allowed us to:

- understand how the bleeding disorders are inherited;
- diagnose carriers of haemophilia;
- sex a fetus very early in pregnancy;
- diagnose haemophilia in the womb; and
- change the instructions in sheep cells so that the animals produce milk containing factor IX.

It is now possible to diagnose haemophilia from a single cell taken before implantation of an embryo into the womb, thus allowing prospective parents the choice of non-haemophilic children. One day it should also be possible to both identify and change a gene defect to remove the possibility of haemophilia in an individual embryo. Gene therapy of this sort may eventually allow us to cure bleeding disorders later in life as well (see Chapter 17).

Within the reproductive organs of testicles and ovaries, the chromosomes in the cells making sperms and eggs divide in a special way. Twenty-three of the chromosomes (one of each pair) go into one new sperm (or egg) and 23 (the second of each pair) into the other. One of the 23 is a sex chromosome, the others are autosomes. This is shown in Fig. 4.9.

Because a man has X and Y sex chromosomes, each of his sperms contains either an X or a Y.

Because a women has two X chromosomes each of her eggs must contain an X.

At conception, an X egg will be fertilized either by an X sperm or by a Y sperm. If the sperm contains an X chromosome the baby will have two Xs (XX) and will be a daughter. If the sperm contains a Y chromosome the baby will have an X and a Y chromosome (XY) and

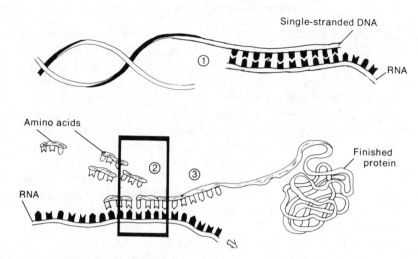

Fig. 4.8. Translation of the genetic code. (1) Instructions on a single strand of DNA are copied exactly by RNA. (2) The RNA is run through a ribosome, together with amino acids derived by the cell from food. Within the ribosome these amino acids are lined up to fit the exact sequence of codons on the RNA. (3) The assembled amino acids leave the ribosome as a length of protein which then folds upon itself to occupy a tiny space.

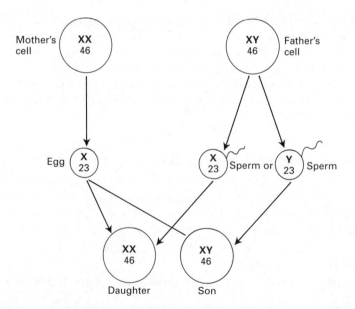

Fig. 4.9. How the sex of a baby is determined. Cell division in the mother's ovaries results in eggs containing 23 chromosomes, one of which is always an X chromosome. Similar division in the father's testicles results in sperm with 23 chromosomes, one of which is always an X or a Y.

Mother   Father   Possible children
                         Daughters   Sons

XX + **XY** = **XX** or **XX** or **XY** or **XY**
1 2   3 4   1 3   2 3   1 4   2 4

Fig. 4.10. The possible combinations of the sex chromosomes.

will be a son. Thus it is the sperm from the father that determines the sex of the child. The possible combination of the sex chromosomes can easily be worked out (Fig. 4.10).

That this makes sense is obvious: there is a 50:50 chance of having a boy or a girl, and, allowing for the chances of survival, the two sexes are about equal in the population.

Genes
Pairs of genes

In giving two sets of 23 chromosomes to the new baby at conception the mother and the father are really giving two separate sets of genetic instructions. Just as the chromosomes are paired, so are the genes. Each human body cell contains 100 000 genes—half from the father and half from the mother. Each gene represents one piece of information.

As an example of how paired genes work, let us imagine a gene which can exist as either ● or •. The possible combinations of ● and • are:

● ●, or ● •, or • •.

Now let us say that the effect of a ● gene on the baby is to give it a big nose and the effect of a • gene is a small nose. The baby with ● from the father and ● from the mother (● ●) will have a big nose. A • • baby will have a small nose. What about ● •?

If the baby with ● • has a big nose we say that ● is a dominant gene—it 'dominates'. In this example • is said to be recessive, because when paired with the dominant gene it has no detectable effect.

How genes work

Genes are made of a substance called deoxyribonucleic acid (DNA). DNA exists as a double helix. It passes its information from the nucleus into the cell factory using a messenger called RNA (ribonucleic acid). From here it is used as the blueprint from which proteins are built. In this way genetic information produces an effect or, in the terms of our cell factory, an end-product.

Most of the blood clotting factors are proteins. They are built in specialized cells from the instructions carried by DNA. These instructions are in the form of a code known as the genetic code. Codes are a sort of language and when the words of a language are written down misprints sometimes occur. For instance, the word DOG might appear as ODG, a meaningless jumble of letters. If this happens to a piece of DNA the protein made from the instructions will be faulty, and will not be able to function properly in the body. The instructions may be so mixed up that the cell factory is unable to produce a protein at all.

A genetic change in which a piece of DNA is altered or lost is called a mutation. Mutations are occurring continually and this is not really surprising in view of all the genes involved in the animal and plant kingdoms. The effects of mutation may be so severe that life is impossible, or so mild that they go unnoticed. They may also produce good as well as bad changes. Without them there would be no change and therefore no evolution.

Once a mutation has occurred, it is permanent. The misprinted gene is fixed to its chromosome and copied every time a cell divides. The misprint may appear in the sperm or egg to be passed to the next generation. If this happens and its effect is to produce a disorder, we say the disorder is inherited, or hereditary, and we expect to find a family history of the disorder. But not always! About 70 per cent of people with haemophilia have a family history of the disorder. The remaining 30 per cent are assumed to be the result of a recent mutation.

If a mutation affects one of the genes governing the production of a blood clotting factor three things may happen:

1. No clotting factor at all is produced.
2. An abnormal (misprinted) clotting factor is produced.
3. Only a small amount of the usual clotting factor is produced.

If no clotting factor is produced there can be no activity in the blood.

If an abnormal clotting factor is produced, it will not work properly and there will be no activity in the blood.

If only a small amount of the factor is produced, the activity in the blood will be less than usual.

These rules apply to the haemophilia factors and to the other protein factors as well. Their recognition resulted in the development of one of the methods for the detection of carriers.

Carriers

Carriers are people who 'carry' an abnormal gene which may produce an effect in their descendants. The knowledge that a woman is a carrier of, say, haemophilia often brings her a feeling of remorse or even guilt. This is, of course, irrational because *every* human carries abnormal genes. It cannot be anyone's 'fault' if the effect of one of these genes becomes apparent in a child. However irrational it is the feelings can be very intense. The best thing to do is to share them with the family and talk about them openly with the centre staff.

Carriers of haemophilia are usually unaware of the genetic abnormality because the other gene in the pair protects them from the abnormal effect. However, this protection is occasionally incomplete and some carriers have clotting factor levels low enough to cause problems with bleeding, especially on surgical challenge. Thanks to genetic engineering techniques the identification of carriers is becoming easier, although some may be identified simply by the history of the disorder in their families, or by the simple measurement of the relevant factor clotting activity in their blood. Whatever the method used it is always wise to have this activity measured in childhood so that if it is low protection can be provided in case serious injury or the need for operation arise. Whilst it may be easy to diagnose whether or not a girl is a haemophilia carrier, it can be very difficult for her parents to tell her. I adapted the following suggestions from an excellent leaflet published by the Haemophilia Foundation of Australia, which was written to help families over the hurdle:

- Consider the readiness of your daughter to deal with information about carrier status by taking into account her age, emotional maturity, and her level of understanding and interest in the information.
- It will only be in adolescent years that abstract concepts will be understood. It may be useful to find out when genetics is taught at school and use it as a background for additional knowledge.
- Information should be given on several occasions over the years. The way that children and adolescents think changes over the years, and the situation needs to be talked through at different levels over time.
- When explaining, use language your daughter can easily relate to. Adopt words she might use herself, so that she can understand more readily and feel more secure.
- Remember that adolescents feel sensitive about their personal image and undergo rapidly changing emotional states. Allow for strong reactions and be supportive (sometimes easier said than done!).
- The attitude of fathers or brothers with haemophilia to their disorder usually provides a model for the carrier daughter's attitude to being a carrier and to the possibility of having a son with haemophilia. A degree of acceptance and open communication between those in the family who have haemophilia and your daughter will help her to accept and deal with her carrier status.

- Ensure that your daughter has access to information and materials so that she can process information about the carrier status in her own terms and therefore feel more secure about it.
- Remember that denial is a way of dealing with carrier status. At the time of formal carrier testing in adolescence, a girl may learn for the first time that she is a carrier. Even if she has known that she may be or is a carrier, it is in adolescence that the implications will be more fully understood.
- Your daughter's anxiety may be made more acute by watching the problems experienced by a brother with haemophilia. This anxiety may be further compounded by feelings of anger towards her brother who sometimes requires special attention for a condition which not only causes her to be ignored, but which she herself may pass on to her children.
- A refusal to discuss anything to do with being a carrier may be an initial reaction. There may be avoidance or denial of the facts or, by contrast, an obsession with the facts which is indicated by constant discussion and reviewing of them. Reaction could be in the form of rebellious or emotional withdrawal or depression. If any of those coping mechanisms persist, parents should seek advice from their family doctor or centre staff.

The bleeding disorders

We are now in a position to understand how the bleeding disorders are inherited. They are divided into **sex-linked recessive, autosomal dominant**, and **autosomal recessive** varieties.

Sex-linked recessive

These include:

- factor VIII deficiency (haemophilia A)
- factor IX deficiency (haemophilia B; Christmas disease).

In these disorders the abnormal gene is on an X sex chromosome. Males are affected because they have only one X chromosome. Their other sex chromosome is the Y, which does not carry a duplicate factor VIII gene. Females are usually normal because their second X chromosome carries a normal gene. The possible children of a carrier of haemophilia A or B are shown in Fig. 4.11.

If ⊗ is the abnormal gene, the chances are one in two (50:50) that each daughter will be a carrier, and one in two (50:50) that each son will be affected.

The chances for any son or daughter inheriting the abnormal gene are those of tossing a coin: heads an affected son, tails a normal son. Heads a carrier daughter, tails a normal daughter. A family may be lucky and

Possible children

Mother	Father	Daughters		Sons	

✃X + XY = ✃X or XX or ✃Y or XY
1 2　　3 4　　1 3　　2 3　　1 4　　2 4

　　　　　　　　Carrier　Normal　Haemophiliac　Normal

✃ denotes sex chromosome with abnormal gene

Fig. 4.11. The possible children of a carrier of haemophilia A or B.

none of the children may be affected. Conversely, all may inherit the defective gene.

A good example of a family tree illustrating the inheritance of one of the haemophilias is that of Queen Victoria (see Fig. 17.1 and p. 344). We will probably never know whether the bleeding disorder that afflicted the Queen's descendants was haemophilia A or haemophilia B, because both are inherited in the same way, and both may have the same effects. In the tree you will notice that the abnormal gene is handed down from the affected father (Leopold) to his daughter. This is because the father *must* give an X to create a daughter. A Y sperm would result in a son (Fig. 4.12).

All the daughters of a father with haemophilia will be carriers. All his sons will be normal, and cannot pass haemophilia on (Fig. 4.13).

Effects

In the case of the haemophilias we know that certain members of a family must be carriers. They are:

(1) the mothers with one haemophilic son and a previous family history of the disorder;
(2) the mothers of more than one haemophilic son;
(3) all the daughters of a haemophilic father.

There will be doubt in the case of:

1. *Families with no previous history and one haemophilic son*
   In this case the solitary affected boy with haemophilia is probably

Possible children

Mother	Father	Daughters		Sons	

XX + ✃Y = X✃ or X✃ or XY or XY
1 2　　3 4　　1 3　　2 3　　1 4　　2 4

　　　　　　　Carrier　Carrier　Normal　Normal

✃ denotes sex chromosome with abnormal gene

Fig. 4.12. The abnormal gene is handed down from the father to all his daughters.

the result of a mutation. The mutation may have arisen only in the fertilized egg, in which case no other children will be affected. However, it may have arisen in the sperm of the mother's father, or the egg of the mother's mother or grandmother, in which case the mother herself will be a carrier, and further children will be at risk.

2. *Daughters of known haemophilia carriers*
   The daughters of known haemophilia carriers have a 50:50 chance of being carriers themselves.

Some of the carriers of haemophilia A or B may be detected by laboratory tests before they have children. The reader will remember that women have two X chromosomes and that the factor VIII or factor IX gene on the second X makes up for the defect on its sister gene. This is only partly true because the overall activity in any one individual depends on how many of the normal genes are switched on. If most of them are working the clotting factor level will be in the normal range. In contrast, if most of them are switched off whilst the abnormal genes are switched on, the level will be low. Some one-third of haemophilia carriers have a lower than normal factor VIII or factor IX activity in their blood. In a few cases this activity is low enough to cause symptoms of the disorder, usually in the form of easy bruising or bleeding after operations or childbirth. If the activity level is below normal in a suspected carrier, it is almost certain that she does carry the abnormality, and that future children are at risk.

Another way of diagnosing haemophilia carriers depends on measuring both the amount of factor VIII or factor IX protein present in the

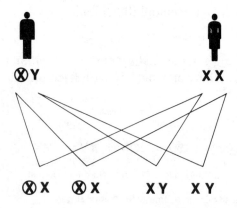

Fig. 4.13. Another way of showing why a haemophiliac's daughters must be carriers, and his sons normal and unable to pass on the infected gene.

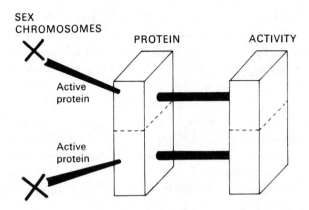

Fig. 4.14. Diagram to show production of active clotting factor in a normal woman.

woman and the level of clotting activity. If the right amount of protein is present for the level of activity then the woman is probably not a carrier (Fig. 4.14). If there is more protein than activity it means that an inactive protein is being produced and this must come from the abnormal gene on one X chromosome (Fig. 4.15). The other X chromosome is producing protein with a corresponding clotting activity. In this case the woman will be a carrier. It should be noted that the result of these tests vary with age, ABO blood group, pregnancy, and the use of oral contraceptives.

Using these tests it is easier to predict carriers of haemophilia A than it is to predict carriers of haemophilia B. This is because the character of haemophilia B differs widely between families and the tests which

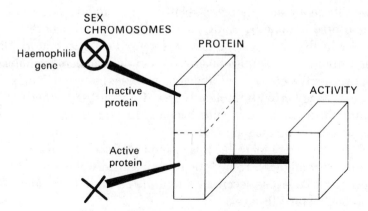

Fig. 4.15. Diagram to show production of active and inactive clotting factor in a haemophilia carrier.

detect the factor IX protein do not do so in a substantial number of patients. Using available antibody to factor IX, which should react with the protein antigen, it is possible to identify some haemophilia B (Christmas disease) patients in much the same way as the protein detection works in haemophilia A. These patients have plenty of the IX protein but it is inactive. Another name for this protein is cross-reacting material, or CRM, and patients in whom the protein can be identified in normal amounts are called CRM-positive (CRM+). They account for about 15 per cent of people with haemophilia B. A further 25 per cent of patients have reduced levels of the protein (CRM^R). In the remaining 60 per cent of patients no protein can be identified by available techniques; these people are termed CRM-negative (CRM−).

It follows that, in attempting to advise possible haemophilia B carriers, it is very important to test other members of the family to see to which group they belong. Detection is just like in haemophilia A in the case of CRM+ families, but much less easy in CRM^R and CRM− families.

Both haemophilia A and haemophilia B carriers can now be identified using genetic techniques. In order to identify the defect several members of a family may need to be tested, including those with haemophilia. All that is required is a small specimen of blood which contains white cells from which DNA can be extracted and stored. These techniques, together with the family history and measurements of clotting activity and related protein, allow for the accurate diagnosis of carriers in over 95 per cent of cases. While it is often possible to locate the precise genetic defect in an individual with haemophilia, this is not usually necessary when trying to identify carriers. An easier way of doing this is called gene tracking.

What a female who may be a carrier really needs to know is whether she has inherited the X chromosome with the gene that caused haemophilia in a male relative or relatives. The identification of such a chromosome is done by using our knowledge of normal variations that occur in the chromosome near the gene. In the laboratory DNA obtained from blood samples taken from family members is analysed. The DNA from the area around the gene makes a pattern of bands that can be read together with the family tree. If the girl has inherited the band present in her relative with haemophilia she must be a carrier. This is shown in the diagram (Fig. 4.16).

In the diagram a mother called Susan and her husband James have two children, Peter and Mary. There is no previous history of haemophilia in the family, but Peter has severe haemophilia A. Both Susan and Mary have normal factor VIII levels.

Beneath the family tree is the result of DNA banding. Reading from left to right, Susan has two bands, 1 and 4, which represent

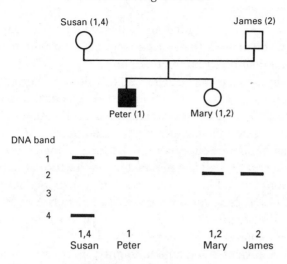

Fig. 4.16. Gene tracking

her two X chromosomes. Peter, who has of course inherited his X chromosome from his mother, has band 1. Mary has bands 1 and 2. The X chromosome from her father James produces band 2.

Mary has therefore inherited band 1 from her mother, and band 2 from her father. Band 1 represents the chromosome present in her haemophilic brother and Mary is therefore a carrier. If she had inherited band 4 from her mother she could not have been a carrier.

There is another possibility in this family. A genetic change (mutation) might have taken place in the egg from which Peter developed, and caused his haemophilia. Such a change would not be part of Susan's genetic makeup and would not therefore show up on the DNA band pattern. A mutation like this is unlikely because we know from research that 90 per cent of the mothers of isolated sons with haemophilia like Peter are carriers. If Mary wants more information on the probability of her being a carrier it may be possible to use other techniques to define the precise genetic defect in Peter, and then to look for it in her DNA.

Gene tracking is much easier when there is more than one person with haemophilia in a family. For instance, if Mary had another haemophilic brother like Peter the identification of band 1 in her would confirm her carrier status.

Until recently the high prediction rate achieved for haemophilia B carriers using DNA analysis could not be matched for haemophilia A carriers. This was because, unlike in haemophilia B, the expected number of individual mutations could not be found. Indeed, in about 50 per cent of cases the genetic cause of haemophilia A remained unknown. However, in 1993 an unexpected change in the factor VIII gene was discovered. It is called an 'inversion' and has been described as a 'flip–tip' change to the normal sequence by which the genetic instructions of factor VIII are read in the cell.

What has happened is that, during cell division, the normal factor VIII gene has doubled up on itself. In doing so an area near one end, or tip, of the gene has come to lie next to an area near the other end. The two sequences of instructions have then swapped places, or 'recombined'. Once this has happened the instructions for making active factor VIII are scrambled, and the cell can no longer read the normal sequence. The term 'inversion' is used because a part of this sequence is spliced into the new, abnormal gene the wrong way round during recombination. The effect is the same as splicing a sequence of sound or vision into an audio or videotape backwards.

A great advantage of being able to identify an inversion in the factor VIII gene is that only the DNA from the person being tested is needed for the analysis. This means that a substantial number of possible carriers from families in which the person with haemophilia has died and from whom DNA is not available, can be identified.

Before leaving our discussion on carrier detection it is important to understand that while nature usually follows the rules and allows us to identify most carriers with great accuracy, very occasionally a hiccough occurs. Even with the most up-to-date techniques it may not be possible to identify every carrier with the precision we would like.

Carrier detection for all the clotting disorders is a time-consuming and expert procedure, and the counselling which follows also requires an intimate knowledge of the family and the relationship between its members.

As was suggested earlier, it is always worthwhile checking the activity level at least once in both obligatory and possible carriers in childhood because, if it is very low, special treatment will be needed in the event of an accident or the need for surgery. More accurate testing can then be performed in adolescence.

Autosomal dominant

These disorders include:

- hereditary haemorrhagic telangiectasia
- von Willebrand disease (most families).

In these disorders one parent carries the abnormal gene, and because it is dominant this parent will show the effects of the gene. Because the gene is on an autosome it can be carried by father or mother and passed on to sons or daughters. What combinations are possible? These are shown in Figs. 4.17 and 4.18.

If ● is the abnormal gene, and the combinations shown are possible children, it follows that there is a two in four (one in two or 50:50) chance that each child will inherit the abnormality.

In other words, the chances are those of tossing a coin; 'heads' the abnormal gene, 'tails' the normal one.

The chance is the same for each child, but just as when a coin is tossed a run of heads or tails may come up, so it is with children. The first may be affected and the second normal, or both may be affected, or both normal, and so on.

The family tree of a person with dominantly inherited von Willebrand disease is illustrated in Fig. 4.19.

Effects

In hereditary haemorrhagic telangiectasia and most cases of von Willebrand disease, the abnormal gene dominates its partner and anyone carrying it will show the effects. However, these are very

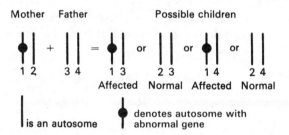

Fig. 4.17. One way of showing how an autosomal dominant characteristic is inherited.

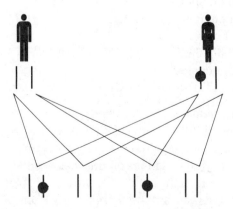

Fig. 4.18. Another way of showing how a dominant characteristic is inherited.

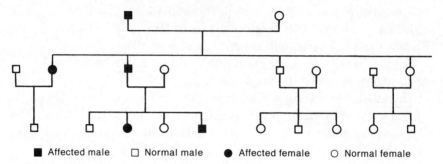

Affected male □ Normal male ● Affected female ○ Normal female

Fig. 4.19. Family tree of person with dominantly inherited von Willebrand disease.

variable from person to person and generation to generation. For instance, a woman with a telangiectactic lesion in the nose may need frequent treatment for nosebleeds, whilst her sister may only have one or two spots on the lips. A man with von Willebrand disease may have no trouble at all, whilst his sister has considerable trouble with her periods. Another problem with these conditions is that they may only show themselves with advancing age. Children rarely have troublesome lesions of telangiectasia and people with von Willebrand disease may reveal themselves only because of problems with menstruation.

Autosomal recessive

These disorders include:

- factor I deficiency
- factor II deficiency
- factor V deficiency
- factor VII deficiency
- factor X deficiency
- factor XI deficiency
- factor XII deficiency
- factor XIII deficiency
- von Willebrand disease (sometimes)

Here both parents must carry the abnormal gene for the disorder to be apparent in the child. Usually neither parent will show a bleeding disorder because each has a paired normal gene which exerts a dominant effect. However, there are variations on this theme, and they are discussed below. The disorder affects the child with two abnormal genes. Once again, because the gene is on an autosome, both sexes can be affected. What combinations are possible? These are shown in Figs 4.20 and 4.21.

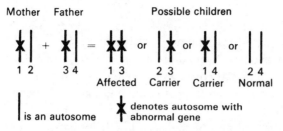

Fig. 4.20. One way of showing how an autosomal recessive characteristic is inherited; both parents carry the abnormal gene.

If X is the abnormal gene, only one child out of four will be affected; the chances for each child are one in four. Once again we cannot say which child in a series will be affected; it might be the first, second, third, and so on. If parents carrying abnormal genes are unfortunate, all their children could be affected; if fortunate, none.

The chances of two people with the same abnormal gene marrying are usually very remote, and, therefore, the disorders produced are rare.

The family tree of a person (arrowed) with factor V deficiency is shown in Fig. 4.22. None of her children is affected; this is because she can only pass one of her abnormal factor V genes to each child, the pair being made up by a normal gene from her husband. It will be noted that her parents are first cousins. When members of the same family marry, the chances of both partners carrying the abnormal gene are greater than in the normal population. This is why the laws governing many peoples forbid intrafamily marriages, and why remote tribes or isolated sects, in which interbreeding has occurred, contain more hereditary diseases than usual. Marriage between relatives is called consanguinity, literally 'with one's own blood'.

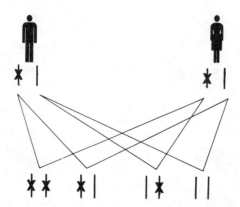

Fig. 4.21. Another way of showing how a recessive characteristic is inherited.

Effects

These only come to light when a child is affected by a disorder known by experience to be inherited in the recessive manner. Both parents must be carriers, and usually only when an affected child appears can they be told that there is a one in four chance of each subsequent child being affected.

The children of an affected adult will be carriers, but the chances of them marrying another carrier are usually very remote. From the number of times a recessive condition appears in the population the number of abnormal genes can be calculated (the gene frequency). This frequency is known for a considerable number of diseases. A family worried about the chances of the disorder appearing in their relatives or descendants should always seek the advice of a specialist in genetic counselling, through their haemophilia centre.

Homozygotes and heterozygotes

The child who inherits two abnormal genes is called a homozygote. As explained above he or she is bound to have the factor deficiency carried by each parent.

Each of his parents is a heterozygote, with one normal and one abnormal gene for the particular clotting factor. We now know that

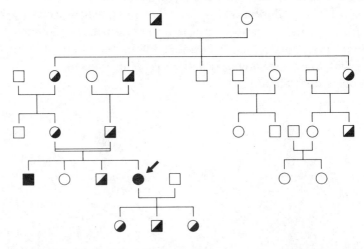

■ Affected male ● Affected female ◪ Carrier male ◓ Carrier female
□ Normal male ○ Normal female ⊏⊐ First-cousin marriage

Fig. 4.22. Family tree of a person (arrowed) with factor V deficiency. The mother and son on the right hand side were identified as carriers by their lower than normal factor V levels.

someone who is a heterozygote may have abnormal bleeding and therefore be at risk after injury or during surgery. The situation is just like that of the haemophilia A or B carrier in whom the factor VIII or IX clotting activity is low. It follows that all the relatives of someone found to have an autosomal recessive clotting disorder should be tested. The reasons why a bleeding disorder may affect hetero-as well as homozygotes are to do with specific genetic mutations which occur in some families, as well as the action of the normal gene. If this action is not a robust one insufficient active clotting factor may be produced to correct the defect caused by the other, inactive gene.

The link between haemophilia A and von Willebrand disease

Although haemophilia A and von Willebrand disease are inherited in different ways, a deficiency of factor VIII is common to both disorders. How can this be explained?

The first clue is given by the different patterns of inheritance. Haemophilia A is a sex-linked disorder; the abnormal gene is sited on the X chromosome. Von Willebrand disease is an autosomal disorder; the abnormal gene is sited on an autosome. Because both conditions result in an abnormality of the same factor, both the X-linked and the autosome-linked genes must play a part in its production.

The second clue comes from the results of transfusion experiments in people with haemophilia A or von Willebrand disease. In the person with haemophilia the level of factor VIII activity rises rapidly after a transfusion of fresh normal plasma. After reaching a peak the activity falls rapidly because of the 12-hour half-life of factor VIII (see p. 47 and Fig. 4.23).

In someone with von Willebrand disease the level of factor VIII activity rises rapidly after transfusion in exactly the same way as in the person with haemophilia. There is then a second, slower rise in the person with von Willebrand disease (Fig. 4.24). After reaching a peak the activity starts to fall away. The rate of this fall is much slower than in the person with haemophilia. This suggests that, in addition to complete factor VIII, fresh normal plasma contains components from which the person with von Willebrand disease can make his own factor VIII.

The person with von Willebrand disease also has a platelet disorder; it is this that is often responsible for the long bleeding time. In many people with von Willebrand disease the bleeding time shortens after transfusion and this suggests that there is something else in the plasma which directly affects the platelets and the way they stick to the blood vessel wall. Work on the structure of the factor VIII molecule has

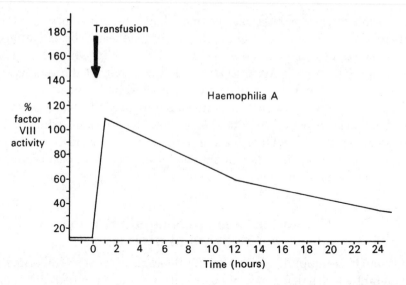

Fig. 4.23. The effect of transfusion (arrowed) in someone with haemophilia A. After a peak the activity falls rapidly because of the 12-hour half-life of factor VIII clotting activity.

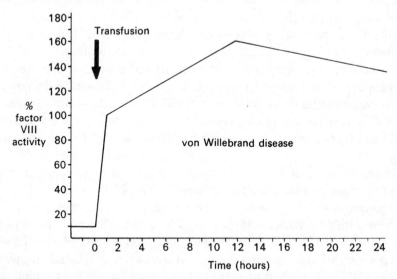

Fig. 4.24. The result of transfusion in someone with von Willebrand disease (for explanation see text).

Table 4.1 Parts of the factor VIII assembly

VIIIC is factor VIII clotting activity
VIIICAg is factor VIII clotting antigen
VWF is von Willebrand factor
VWFAg is von Willebrand factor antigen.

Von Willebrand factor is made of a number of building blocks called multimers. These can be separated out and identified in the laboratory. Diagnosis of the type of von Willebrand disease someone has depends in part on this test, which is called multimeric analysis. Diagnosis also depends on measuring the activity of VWF which is reflected in the bleeding time and on how the platelets stick together in the presence of an antibiotic called ristocetin.

RICOF is this ristocetin cofactor.

---

revealed that the prolongation of the bleeding time is related directly to the von Willebrand part of the assembly.

The two parts of the factor VIII assembly in the bloodstream are called VIII:vWF (VIII: von Willebrand factor) and VIII:C (VIII: Clotting). VIII:vWF is made in the lining of the blood vessel walls. VIII:C is made in the liver. In order to function properly VIII:vWF and VIII:C have to link up. If either component is deficient or faulty, or if the link-up does not occur, a bleeding disorder results.

New laboratory tests are helping to identify parts of the assembly, and these have been given names and shorthand descriptions. They are shown in Table 4.1.

The factor VIII molecule can be represented as in Fig. 4.25.

In severe von Willebrand's disease, VIII:vWF is missing, and because we think that VIII:vWF is necessary to protect the delicate clotting activity VIII:C, that cannot function either.

We can now explain the effect of transfusion (p. 84) in another way. When plasma is given to someone with haemophilia A only the VIII:C pieces are useful; once they are used up the effect of the transfusion is over. When plasma is given to someone with von Willebrand disease both VIII:vWF and VIII:C pieces are useful. The VIII:C pieces produce an immediate rise in clotting activity and the VIII:vWF pieces start to protect the VIII:C, which the person with von Willebrand disease is perfectly capable of making. Hence the secondary rise in clotting

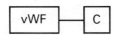

Fig. 4.25. Representation of factor VIII molecule (vWF, von Willebrand factor; C, clotting factor).

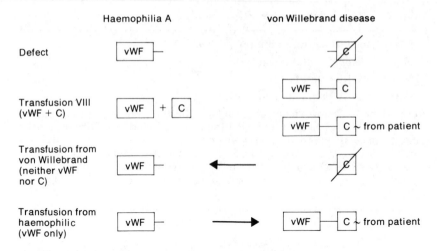

Fig. 4.26. Transfusion experiments in severe haemophilia A and severe von Willebrand disease. In the von Willebrand plasma the VIIIC is unprotected.

activity shown on p. 84 (Fig. 4.24). That this actually happens is clinched by showing that when von Willebrand plasma is transfused into someone with haemophilia A nothing happens (Fig. 4.26). When plasma from someone with severe haemophilia A is transfused into someone with von Willebrand disease there is a rise in VIII:clotting activity. The haemophilic plasma can only have provided the VIII:vWF pieces needed to allow the system to work.

Varieties of von Willebrand disease

As a result of our increasing knowledge of the structure and inheritance of the factor VIII: von Willebrand molecule, three main types of von Willebrand disease are now recognized (Table 4.2). Further subdivision into a number of categories has recently been proposed (Table 4.3). Although complex and reflecting our imperfect understanding of the complete picture, these categories are important clinically because the treatment of the bleeding disorder they cause may vary. In order to decide the best treatment for an individual the response to an injection of desmopressin may be required. This might also be necessary when trying to sort out the inheritance of von Willebrand disease in a particular family.

Table 4.2 The broad definition of von Willebrand disease (VWF = von Willebrand factor)

Type	Defect	Inheritance
1	Partial deficiency VWF	Often dominant
2	Abnormal VWF	Dominant or recessive
3	Absence of VWF	Recessive

Table 4.3 A guide to the present (1994) subcategories of von Willebrand disease (VWD), and their treatment. Sometimes measurement of response to desmopressin (DDAVP) is needed in order to determine which treatment is best for a particular person with VWD. A factor VIII concentrate containing enough von Willebrand factor (VWF) to be effective is required when desmopressin is contraindicated. In the list HMW = high molecular weight. The M of 2M stands for 'multimer', and the N of 2N for 'Normandy'. Further explanation is in the text.

Type and subtype	Characteristics
1	Should respond well to DDAVP.
2A	Absence of HMW multimers together with defective platelet interaction. Should respond to DDAVP.
2B	Defective platelet interaction. DDAVP usually contraindicated and VWF replacement therapy used instead.
2M	Abnormal VWF multimer structure associated with abnormal platelet interaction. Should respond to DDAVP.
2N	Abnormal affinity for factor VIII. VWF replacement therapy required.
3	VWF replacement therapy required.

PART III

Treatment

5

When and what to treat: bleeding episodes

Haemophilia A and haemophilia B (Christmas disease)

The effects of these disorders are very similar and they will be discussed together. They are related to the clotting activity level of factor VIII in the blood in the case of haemophilia A, and of factor IX in haemophilia B. They are also related to the individual. Some people with haemophilia with very low factor levels rarely experience trouble, whilst others with the same level have many bleeds. Whilst most people with haemophilia A have roughly the same degree of factor VIII deficiency throughout life, levels of factor IX rise steadily with age in some of those with haemophilia B (Christmas disease). I suspect that there are some families with mild haemophilia A in which factor VIII does the same.

If there is a family history of the disorder it is usually found that different members of the family are affected to the same extent, both in terms of factor level and in terms of frequency of bleeding. Some people recognize phasic or seasonal bleeding, experiencing their most severe episodes in the Spring and Autumn (or Fall). Some seem to bleed more when they are tense and anxious.

The descriptions of the different episodes in this chapter refer to those which might be found in the most severely affected person. *It is very important to realize that the bleeds described are not all going to happen in any one person, and that all can be alleviated with proper treatment.*

Bleeding in the haemophilias is typically prolonged. It is no faster than would occur in someone without haemophilia who has the same injury. Contrary to popular belief, people with haemophilia do not rapidly collapse in pools of blood. Therefore, the same first aid measures used to help anyone who has been injured are just as relevant for someone with haemophilia.

Open bleeds
Minor injuries to the skin and membranes

Because blood-vessel contraction and platelet function are normal, pin-pricks and small wounds seal off quickly and rarely give any trouble. Exceptions can be cuts involving the tongue and lining of the mouth, and around the scalp and face. Small tongue bites and tears of the frenum (the ridge of tissue which projects downwards in the mid-line between upper lip and gum) may result in persistent oozing. This is because these tissues are very rich in blood vessels and are constantly being disturbed when eating, drinking, or talking. A rich blood vasculature is present in the face and scalp, and these areas are also subject to very frequent movement which is likely to disturb the formation of platelet plugs and early clots. Open razor nicks cut straight through small vessels cleanly with little tissue damage and this, and the presence of oils from shaving soap, results in rather prolonged bleeding even in normal people. Many adults with haemophilia use an electric shaver.

Other injuries to the skin and membranes

Deep cuts anywhere in the body will continue to ooze. In the mouth, persistent bleeding from the gums may be the result of infection caused by poor dental hygiene, and dental abscesses will cause pain and bleeding. Children sometimes injure the roof of the mouth on lollipop sticks, and the inside of the nose and ear with sharp objects like pencils and sticks. The lower part of the septum (the division between the nostrils) of the nose has many blood vessels and these are sometimes ruptured in nose-picking. When this happens fingernails may contain traces of blood. Nosebleeds are common during colds, or may result from small growths called polyps (which are not cancers) and occur in people without haemophilia. When nosebleeds (epistaxes) are frequent, note which nostril the blood comes from and tell the doctor. Bleeding in or around the eye should always be treated very seriously.

Bleeding from within

The coughing up of blood from the lungs is not a feature of haemophilia. When it occurs it is likely to be the result of some other trouble and should always be reported to the doctor. Children may appear to cough up blood which is actually coming from the back of the nose or from a lesion in the mouth. Severe coughing spells in, for instance, bronchitis or whooping cough lead to an increase in the pressure of blood in the vessels of the head and neck and may precipitate bruising around the eyes or bleeding from the nose or mouth.

Background information
The digestive system

Beginning at the mouth and ending at the anus, the digestive system consists of a continuous tube into which several ducts and glands open (Fig. 5.1). Food is chewed by the teeth, mixed with digestive enzymes (substances that start the breakdown of food into absorbable parts), and tasted by the tongue. The act of swallowing closes off the air passages and opens the gullet to accept the food, which then passes through the muscular gullet (oesophagus) and into the stomach.

The stomach secretes acid and more enzymes which are mixed with the food. When the mixing is complete, portions of food are expelled into the small bowel, the first part of which is called the duodenum. Here they are mixed with more digestive chemicals from the liver, gall-bladder, and pancreas. By now the food is in the form of easily absorbed units of carbohydrates, proteins, and fats. These are absorbed through the walls of the small bowel into the bloodstream and taken to the liver.

Because the acid from the stomach would be harmful to the bowel as a whole, and because some foods require an alkali for breakdown, the acid is neutralized by bile. The stomach and bowel are prevented from digesting themselves by a thin layer of sticky mucus secreted by special glands in their walls.

At the junction of the small and large bowel lies the appendix, a rudimentary organ of use mainly to rabbits who use it to digest grass. With evolution it will disappear from man; until it does it keeps the surgeons busy by becoming inflamed, usually late on a Saturday night!

Within the large bowel, water is absorbed from the remains of the food, and the waste products are then compressed and presented to an expansion of the bowel called the rectum, which lies to the left of the pelvis. Here the stools, or faeces, await expulsion from the body in the act of defaecation. The digestive system behaves like a mass-production line, the food being divided into components necessary for growth and health, which are absorbed, and waste, which is excreted.

As well as carbohydrates, proteins, and fats, the body needs small amounts of vitamins and minerals, like iron, to remain healthy. Advertisers constantly use this knowledge to boost their products and their profits. The average diet of Western man contains far more of all the ingredients needed for health than the body actually needs. Expensive vitamins and 'tonic' preparations go the same way as the money used to pay for them—out! If a person requires a dietary supplement his or her doctor will know; the most usual supplement for people with bleeding disorders is iron.

The liver is protected by the ribs on the right side of the abdomen. It is

Background (cont.)

a vital organ with numerous functions, one of which is the manufacture of clotting factors including factors VIII and IX. It plays an important role in the regulation of many other body constituents, and is concerned with the removal of ammonia, a waste product, from the body. The ammonia is converted into urea, a constituent of urine. The bile produced by the liver contains the waste from the natural breakdown of red blood cells. In liver disease (hepatitis) the level of this breakdown product in the blood rises and produces yellow jaundice. Bile is stored in the gall-bladder, from where it is discharged into the bowel, where it helps in the digestion of fats.

Like the liver, the pancreas also excretes digestive chemicals into the bowel. The other function of this organ is the production of insulin, needed for sugar regulation. If insulin is not produced, sugar diabetes results.

The other abdominal organ is the spleen, which is not concerned with digestion. The spleen lies below the left ribs, opposite the liver. It plays a role in the storage of platelets, and is involved with the protection of the body against infection.

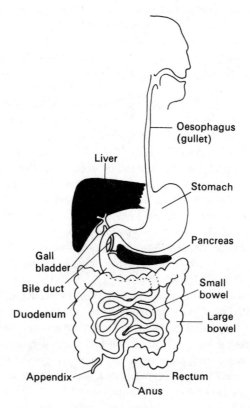

Fig. 5.1. Diagram of the digestive system.

The medical name for the vomiting of blood is 'haematemesis'. In children a haematemesis often occurs when a quantity of blood from the back of the nose has been swallowed—it irritates the stomach and is brought up. Two of the commonest causes of haematemesis in older people are aspirin and alcohol. Both irritate the lining of the stomach, causing acute gastritis. The lining becomes inflamed and suffused with blood, small erosions start and vessels rupture. Alcoholic gastritis may occur when beer or spirits are taken in excess, and typically follows 'a night on the tiles'. Retching after alcohol can also cause the Mallory Weiss syndrome in which a tear occurs at the lower end of the gullet (oesophagus). Haematemesis is also one of the signs of a peptic ulcer.

Bleeding anywhere in the stomach or bowel will invariably result in some of the blood finding its way into the stools (faeces). If it has come from the upper part of the bowel it will mix with the stools and give them a black tarry appearance—though stools will also be made black with iron medicine. Blood from the lower bowel will retain some of its red colour, and the blood from the rectum and anus will streak rather than mix with the stools. There are many reasons for blood in the stools (medical name: melaena) but, apart from those already mentioned, one of the commonest is piles (haemorrhoids). Another common cause, especially in childhood, is constipation when straining at stool may cause a very small tear on the edge of the anus. This must be given a chance to heal or the child will try not to go to the lavatory with the result that further hard stools will exacerbate the problem. The prescription of a mild laxative like Senokot for a week or two softens the stools and relieves the child's pain. A rare cause, also sometimes seen in children, is bleeding from a small pocket of bowel, rather like a second appendix, called Meckel's diverticulum. This structure represents the remains of the opening between the bowel and the yolk sac in the developing baby. In some people a part of the passage persists and may become inflamed and bleed.

Background information
The urinary tract

Urine is formed in the kidneys. The two kidneys lie at the back of the abdomen, one on each side of the spine. They are protected from injury by the lower ribs, spine and back muscles, and the other abdominal organs. The function of the kidneys is to cleanse the blood of waste products. Blood flows through tiny coils of vessels where waste chemicals are filtered into a system of collecting tubes which eventually join up to form the ureters (Fig. 5.2).

A ureter passes from the root of each kidney to join the bladder, which lies in the pelvis. The bladder is a collecting reservoir in which the

Background (cont.)

urine, which is a mixture of water and waste products, is stored until it is passed through the urethra, which runs from the base of the bladder to the outside of the body.

In the male the urethra is joined by the tubes from the testicles in the prostate gland, which lies just below the bladder. It then passes through the penis, with the dual function of conveying either urine or semen. In the female the urethra is much shorter and opens between the lips of the external genitalia, in front of the vagina. A system of muscular valves, controlled by the nervous system, allows for the retention of urine in the bladder until it is convenient to pass it and, in the male, the passage of either urine or semen separately. In both male and female the urethra twists through its course from the bladder to the exterior, which explains why the stream of urine is not a simple jet.

Blood in the water (medical name: haematuria) is a common occurrence in some people with haemophilia. Once again it can be a sign of numerous disorders in other people, but in haemophilia it usually occurs

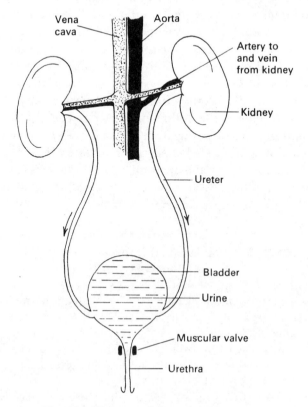

Fig. 5.2. Diagram of the urinary tract.

for no very good reason. Occasionally a colicky pain accompanies the haematuria. It only takes a tiny quantity of blood to stain a lot of urine red, and haematuria always looks far more frightening than it really is. The blood often comes from the kidneys and clears without detectable damage, but it may arise from anywhere in the urinary tract. If fresh blood is passed before the main stream of water the lesion will be found in the lower part of the tract—at the base of the bladder or within the tube through the penis. As is the case elsewhere in the body, bleeding in the urinary tract may follow infection and inflammation. It is important that anyone with blood in their water should *not* take antifibrinolytic medicine like Cyklokapron or Amicar (see p. 137) without careful medical guidance. This is because these drugs may provoke the formation of small clots in the urinary tract and these will cause intense pain. This pain, called renal colic, can only be relieved either by the passage of the clot, or by injections of the pain-killing drug pethidine.

All bleeding from within should be treated seriously and investigated by a doctor. It is usually put right very easily. The important thing is to find the cause, as well as to treat the haemophilia.

Closed bleeds

In this section we will be looking at bleeds into the soft tissues, which include the skin and muscles, into joints and bones, and into the head.

Soft tissues

Easy bruising is, of course, one of the hallmarks of haemophilia, and is noticed first at about the time the infant starts to crawl. The bruising is very variable and its extent depends on the injury causing it and the site of the body affected. Sometimes bruising is noticed and no cause can be remembered; this sort of bleeding is called 'spontaneous'. It is probably very rare for bleeds to just 'happen'. The spontaneous, out-of-the-blue bleed is more likely to be the result of an injury too small to be consciously recognized.

The typical haemophilic bruise feels bumpy. It is raised in the centre and spreads under the skin rather like water-colour paint spreading on a wet canvas. The spreading is a direct result of the failure of haemophilic blood to clot.

Bleeding within the tissues, which is normally brought to a rapid halt by clotting, will continue in the person with haemophilia until

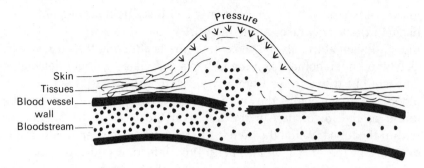

Fig. 5.3. Bleeding in soft tissue in someone with haemophilia.

the pressure of the tissues is equal to the pressure of bleeding from the injured vessels. Because of this, more bleeding will occur in normally lax tissues than in normally tense ones (Fig. 5.3).

Bruising near the surface of the skin is referred to as superficial bruising. It follows knocks or minor injuries like insect bites, and is rarely painful or dangerous. ALL healthy boys (and not a few girls!) have the occasional superficial bruise on their shins or knees from playing football or scrapping.

Sometimes skin bruising indicates trouble deeper in the tissues, because blood from deep structures may track to the surface. Bruising of this sort will be associated with pain, swelling, and sometimes loss of movement. If bruising occurs around the head or neck a careful watch should be kept. Rapid spread, complaints of headache, dizziness or feeling sick, or of difficulty in talking, swallowing, or breathing mean that help should be sought without delay. Similarly, bruising over the kidneys should be reported if it is associated with pain, swelling, or haematuria.

Bleeding may, of course, occur in any of the deeper tissues without recognized injury or visible bruising; it reveals itself with pain, swelling, or loss of function. In infected areas, however, there may be difficulty in deciding whether the signs are due primarily to the inflammation or to bleeding. Mumps is a particular example; others are sore throats or tonsillitis, either of which may provoke bleeding. It is, therefore, a safe rule to seek early treatment for all infections.

Bleeding into muscles

Muscles, and the nerves and blood vessels which run alongside or through them, may be affected either by bleeds within their substance or by bleeds pressing on them from neighbouring tissues. The pressure created in untreated bleeds is sometimes sufficient to cut off the blood supply to many fibres or even the major bulk of a muscle. The result is muscle death, the normally healthy muscle being replaced with scar tissue. As it forms, scar tissue slowly contracts and if it continues this relentless process can pull joints out of true. Drop-foot, in which the person cannot put his heel on the ground when the leg is straight, loss of knee extension, and forms of claw hand are examples of deformities which may occur following severe untreated muscle bleeds.

Muscle bleeds follow blows, sudden twists and sprains, and intra-muscular injections, and often involve a group rather than a single muscle. The person with haemophilia may recognize a cause, or the bleed may be spontaneous. Although all muscle bleeds may cause trouble, the following are of especial importance with regard to early recognition and treatment.

Bleeding into the forearm muscles

These muscles stabilize and work the complex manoeuvres of the wrist, hand, and fingers. Contractures following bleeds may eventually render the whole hand on the affected side virtually useless. Early treatment followed by physiotherapy will maintain good function.

Bleeding into the large muscle masses of back or thigh

Because of their bulk these muscles can accommodate large volumes of blood. As an example, a man fracturing the shaft of his femur in an automobile accident will lose a litre of blood into the tissue surrounding the break. Without transfusion it will take his body two days to adjust his blood volume to normal and at least a month to restore his red cells and haemoglobin. In someone with haemophilia the blood compresses surrounding tissues which may respond by producing a fibrous capsule converting the lesion into a muscle cyst.

Background information
Muscles

Bones and joints provide the frame on which muscles act. The muscles of the skeleton are called voluntary muscles because they can be moved at will.

Voluntary muscles are made up of a multitude of muscle fibres bound together in sheaths. The muscle fibres are capable of contraction and when many of them act in unison a powerful force is created. Voluntary muscle is under nervous control; an electrical impulse causes the contrac-tion. Muscles are usually attached at each end to bone by tendons. They

Background (cont.)
cross one or more joints and exert their force on these joints, producing movement.

As an example of how muscles work, consider the straightening of the knee. The command to do this is coded into electrical impulses in the motor area of the brain. After crossing to the opposite side of the body the impulses run down the nerves in the spine and connect with the femoral nerve which supplies the powerful quadriceps muscle group at the front of the thigh. At its upper end the quadriceps is attached to pelvis and femur, and at its lower end to the tibia, with the kneecap lying in the tendon. The nervous impulses cause the quadriceps to contract and shorten, and the tibia is pulled forward to straighten the knee. As the knee straightens the hamstring muscles behind the thigh gradually relax to produce a smooth, controlled movement (Fig. 5.4).

Muscles are richly endowed with blood vessels (which give them their red colour) and this is why muscle bleeds are common in the severe bleeding disorders.

We talk about the 'tone' and 'power' of muscles, and sportsmen refer to 'getting into shape' by exercising. Muscles work best when they are used regularly—witness the arms of a labourer, the shoulder muscles of an Olympic swimmer, or the calf muscles of a ballet dancer. Unused muscles are weak and flabby. They lack tone, which in medical terminology means the ability of a muscle to develop or resist a force for a long period of time without change in length. Without tone a muscle collapses quickly and, as muscles help strengthen joints, this produces instability. The strength of a muscle is a fickle thing. Disuse rapidly reduces it and much time is required to restore it to its original state. Prolonged bed rest or splinting, or regularly sitting for hours in front of the television set, quickly reduces muscle strength and joint protection.

Bleeding into the psoas and iliacus muscles

Swelling in these muscles (Fig. 5.5) because of bleeding results in pain, loss of skin sensation at the front of the thigh, and a limp. Treatment should be given as soon as possible and will need to be repeated, sometimes for several days in order to prevent early relapse.

The muscles arise from the lower spine and the rear of the pelvis and sweep forward beneath a ligament crossing the top and front of the thigh. In the thigh they are attached to a knob of bone on the inside of the femur. When they contract they help bend the trunk forwards at the hip. One of the major nerves to the lower limb, the femoral, lies between the muscles as they cross the pelvis. This nerve, together with the femoral artery and vein, passes under the ligament

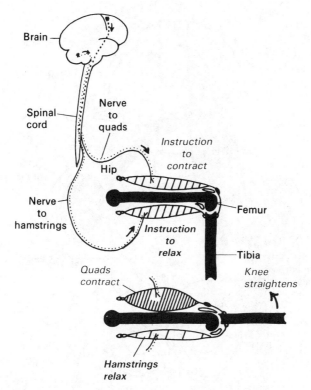

Fig. 5.4. Diagrams to show muscle response (for explanation, see text).

with the muscles. Bleeding into psoas or iliacus will bring pressure to bear on the nerve (Fig. 5.6), particularly in the limited space between the bony pelvis and the ligament. Any nerve responds to pressure by 'going to sleep'—witness the numb feeling in a leg after sitting in one position for along time, and the 'pins-and-needles' which occur as the nerve 'wakes up' and feeling returns. This is exactly what happens to the femoral nerve when it is compressed and, as some of its branches supply the skin over the front of the leg, one of the signs of a psoas/iliacus bleed is loss of sensation in this area.

Distension of the muscles with blood also causes pain. This will be felt in the lower abdomen, and if the bleed is on the right side it may mimic appendicitis. In an attempt to relieve the distension and thus the pain, the person with haemophilia will lie with his thigh flexed at the hip.

Remember that the longer a psoas/iliacus bleed is left untreated, the longer will be the period of recovery. A badly damaged nerve takes many weeks or even months to return to normal.

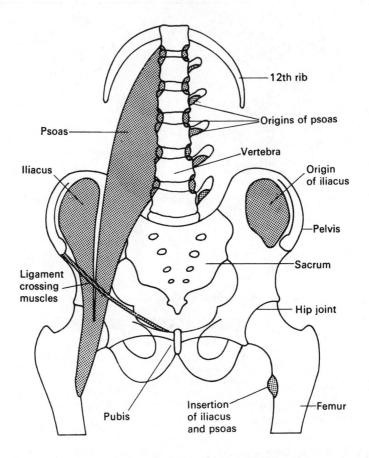

Fig. 5.5. Diagram of psoas and iliacus muscles.

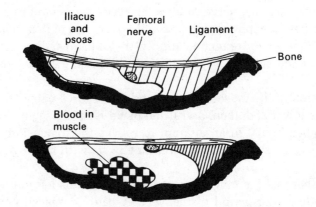

Fig. 5.6. Compression of femoral nerve (hatched area contains another muscle and the femoral artery and vein).

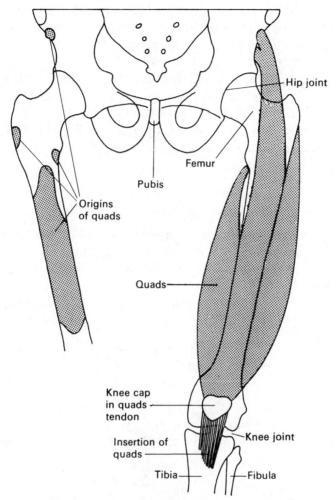

Fig. 5.7. Diagram of the quadriceps groups of muscles.

Bleeding into the quadriceps and hamstrings

The 'quads' are the group of powerful muscles at the front of the thigh (Fig. 5.7). They start from the pelvis and femur and come together in a common tendon which contains the kneecap (patella). The tendon is inserted into the front of the tibia at its upper end. The whole muscle group is a powerful extensor of the knee joint—that is, it straightens the bent knee. It also helps to flex the hip joint.

Although bleeding into the quads is not common, this muscle group is of vital importance to those with haemophilia because of its action on the knee. Weak quads result in an unstable knee which is in danger of recurrent bleeds. Quads power is lost very quickly with rest. A fit professional footballer with normally immensely strong quads may take many days to restore the power after injury.

The hamstrings are the muscles at the back of the thigh, running between pelvis and femur, and tibia. Like the quads they are normally very powerful. Their actions are to flex the knee and extend the hip. Although less likely to cause trouble than the quads they are nevertheless an extremely important group and should not be allowed to weaken by prolonged rest. Untreated hamstring bleeding may cause a flexion deformity of the knee.

Bleeding into the calf muscles

Untreated bleeding into these muscles (Fig. 5.8) will result in a contracture pulling on the Achilles tendon. The heel bone will be pulled up and the person will have to walk on his toes and the ball of his foot.

Intramuscular injections (for instance of penicillin) can provoke nasty

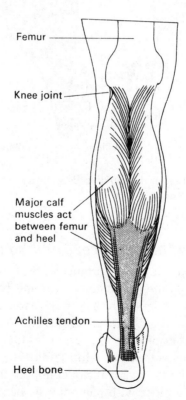

Femur

Knee joint

Major calf muscles act between femur and heel

Achilles tendon

Heel bone

Fig. 5.8. Diagram of the calf muscles.

muscle bleeds and are consequently banned in people with bleeding disorders. The only exceptions to this rule are the small injections required for immunization (p. 7), which are safe, providing pressure is maintained over the site of injection for a few minutes after the procedure.

Bleeding into joints

The first joint bleed (haemarthrosis) occurs in the pre-school years and heralds intermittent bleeding, affecting predominantly knees, elbows, and ankles. Any joint with a synovial membrane may be affected, but it is not by chance that the brunt of the damage should fall on these three joints. All three are basically hinge mechanisms; they are exposed to trivial injury and they are subject to both side-to-side and twisting strains. The knees and ankles must support the weight of the body and the forces involved in the acts of kneeling, sitting, standing, walking, and running. Elbows are often exposed to direct pressure at work. In comparison the ball and socket joints of shoulder and hip are well protected.

An experienced person with haemophilia will know he has started to bleed into a joint before any measurable signs of the haemarthrosis appear. The duration between this feeling and the visible signs may be short or it may be several hours. Whatever the pattern, the sooner treatment is given the less the subsequent damage.

Fig. 5.9 shows the stages through which a synovial joint will pass when subjected to recurrent bleeds. First, blood from a small ruptured vessel in the synovium will flow into the joint cavity. Unchecked, the bleeding will continue until the pressure exerted by the joint capsule and the surrounding tissues is sufficient to shut off the flow. By this time, the bone ends will be forced apart and the capsule distended and tense. The stretching of the capsule causes the pain. The affected

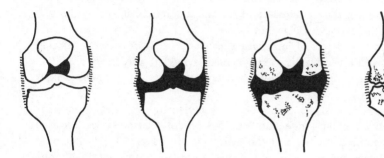

Fig. 5.9. Diagram of long-term progression of joint damage with repeated bleeds.

person will automatically move his joint into the position of maximum comfort—the knee and elbow will be held partly flexed and the ankle partly extended. In these positions the capsules are at their slackest but, of course, continued bleeding will soon take up this slack and acute pain will return with further distension.

Background information
Bones and joints

Bones (Fig. 5.10) are made from calcium, phosphate, and protein. They give support to the body, and protection to the internal organs. The marrow of some bones is the birthplace of many of the blood cells.

The bones of the arm are called the humerus, which lies between shoulder and elbow, and the radius and ulnar, between elbow and wrist. The wrist itself is composed of eight small bones (the carpus) which, together with the metacarpal bones, give the hand great dexterity and mobility. There are 14 phalanges, three for each finger and two for the thumb.

The bones of the lower limb are the femur or thigh bone, and the tibia and fibula, which lie between the knee and ankle. The back half of the foot contains seven bones, the largest being that in the heel. Collectively these are the tarsus, and the metatarsals lie between it and the phalanges of the toes, two for the big toe and three for each of the other toes.

Bones serve to anchor muscles, and the pull of a muscle, particularly if it is powerful and used often, can be seen both on the surface and in the internal structure of the bone at the muscle attachment. The living bone adapts itself to counter the pull and becomes thicker and stronger at this point as a result.

Joints

There are three different sorts of joint in the body. Fibrous joints are simply where two bones are joined by strong fibres and hardly move; the plates of the adult skull are joined in this way. In cartilaginous joints the body surfaces are joined by cartilage—the dense white elastic substance encountered when carving a chicken. Cartilaginous joints all lie in the midline of the body and comprise a joint in the sternum, the discs between the vertebrae, and the pubic joint, which joins the two sides of the pelvis at the front.

It is the third sort of joint which causes trouble in severely affected people with bleeding disorders. Named 'synovial', these joints include those of the shoulders, elbows, wrists and hands, hips, knees, ankles and feet.

The contact surfaces of the bone are called the articular surfaces as they articulate with one another. The articular surfaces are covered with smooth cartilage, and the synovial joint (Fig. 5.11) is enclosed in a fibrous capsule attached near the edges of this cartilage. The non-articular surfaces of the interior of the joint are lined with synovial membrane, or synovium. This delicate membrane contains many small blood vessels, and secretes a thin fluid which acts as a lubricant in the joint, like oil in working machinery. Synovial fluid reduces friction and protects the articular surfaces from wear and tear. It is ironic that such a beautiful system, designed to protect normal joints, should be the cause of much of the crippling associated with untreated haemophilia. Joint bleeds in this disorder start from breaks in the blood vessels of the synovium.

Some joints, for example the knee, are strengthened by ligaments both inside and outside the capsule, giving great strength to the union. Muscles and tendons also serve to strengthen joints. Hence the importance of exercise to maintain a strong, supportive musculature. To reduce friction over moving surfaces, small bags of fluid called bursae (singular: bursa), sometimes communicating with the joint cavity, are found between tendons or ligaments and bones.

The mechanism of a synovial joint depends on its situation and the job it is required to do. The shoulder and hip joints need a wide range of movement and work on the ball and socket principle. The elbow and ankle are hinged, and the knee is a modified hinge joint which is capable of bending, sliding, rocking, and rotating during movement. Other joints, like those in the hands or feet, work by pivoting, rotating, or sliding.

As the joint fills with blood the second stage begins. Red cells break down within the cavity and scavenger cells in the synovium begin to remove the debris. This process has two effects. The synovium becomes laden with iron and other blood waste, and debris remaining in the cavity attacks the smooth surfaces of the cartilage at the ends of the bones, cartilage which is no longer protected by the thin oil of synovial fluid. Bleeding may also start under the cartilage, either as a direct result of the injury or as a result of the pressure from within the joint.

After a series of bleeds in the same joint the third stage is reached. Cartilage is now ragged and pitted and the friction within the joint increased. The synovium is thickened and engorged with waste. Bone ends are softened and ligaments and tendons slackened, partly from disuse and partly from intermittent stretching during the bleeds.

The fourth stage is an extension of the third. Patches of cartilage disappear, leaving bone exposed to the cavity. The softened bone begins to splay out and small areas of dead bone form cysts. The bone ends, without the tension and support of strong ligaments, shift into unnatural positions, and the joint gradually becomes a chronic handicap. The difference between a normal knee joint and the knee of someone with the arthritis of severe haemophilia is shown in Fig. 5.12.

A very important factor in the development of haemophilic arthritis (Fig. 5.13) is muscle wasting, and the vicious circle of joint bleed, prolonged rest with consequent muscle weakness and instability, and a further bleed is known only too well by older people with haemophilia.

Fractures

Bones are more likely to break after long spells of bed rest or confinement to a wheelchair. This is because the muscles are weak and falls are commoner, and because the bones are less tough. Bone is a living structure and is constantly being torn down and built up in the body. This remodelling is due to stresses and strains exerted by muscles; if

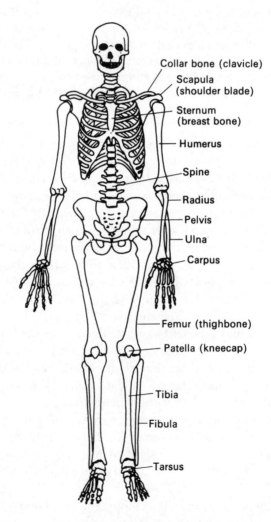

Fig. 5.10. The human skeleton.

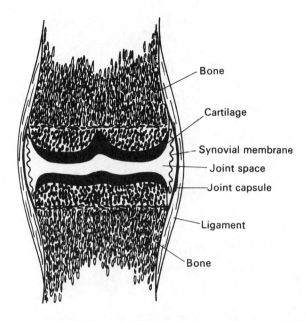

Fig. 5.11. Diagram of a synovial joint.

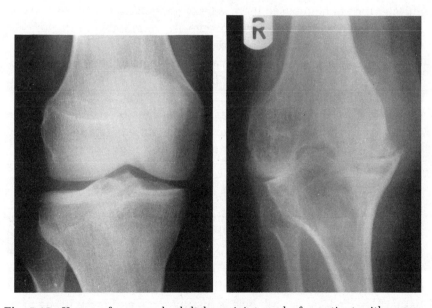

Fig. 5.12. X-rays of a normal adult knee joint, and of a patient with severe haemophilia A who had received inadequate treatment. In the normal joint the 'space' between the femur and tibia is cartilage, through which X-rays pass without leaving an image on the plate. The cartilage in the haemophilic joint has disappeared as a result of recurrent bleeds, so there is no joint space between the bones, which are soft and contain cysts.

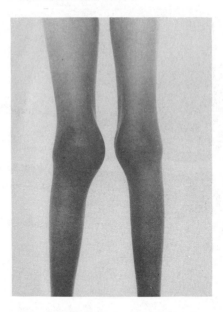

Fig. 5.13. Haemophilic arthritis with muscle wasting.

muscles are not being used there is less remodelling and bones become brittle. This is why old inactive people are more liable to fractures in falls which younger people would shrug off.

Simple, closed fractures rarely cause problems in haemophilia. Provided the broken ends are set properly they will unite in the normal way and in the usual time with remarkably little haemophilia treatment. However, a careful watch is necessary for signs of swelling under a plaster cast. This can indicate the need for further treatment and, of course, the quick removal and refitting of the splint (see p. 36).

Pseudotumours

These very rare lesions are peculiar to the haemophilias, and their cause is not fully understood. One theory is that a small bleed beneath the tough lining of a bone strips the lining away from the bony surface. Bone immediately below the bleed dies and further bleeds in the same place eventually cause cyst formation in adjacent tissues. Over a long period, bleeds disrupt more and more tissue until the lesion presents as an unsightly lump. Hence the name pseudo (false) tumour (growth). Pseudotumours are not cancers. Treatment is usually by surgical removal. There is some evidence that radiotherapy may halt the growth of a pseudotumour if surgery is not possible.

Bleeding into the nervous system

Of all the manifestations of haemophilia, bleeding inside the skull and brain is the most feared. Yet with quick and sustained treatment even this rare complication can often be alleviated.

Background information
The nervous system

The nervous system consists of the brain and a network of nerves which reaches every part of the body via the spinal cord.

The brain is protected from injury by the bony box of the skull, and the main cable of nerves, the spinal cord, is similarly protected by the spine (Fig. 5.14). To allow for movement, the spine is made up of 33 separate bones, the vertebrae, which are jointed together. Within each spinal joint is a pad or disc of tissue which acts as a shock absorber.

Functionally, the brain may be thought of as a remarkably intricate computer, and the nerves as cables which relay commands from the computer and carry information back to it from the body. The power supply of the brain is electrical, and the minute currents needed are created from chemical reactions in the cells. In order to work electrically without shorting, nerves must be insulated from one another, just like the wiring circuit in a house. This insulating process is slow and is incomplete at birth. As it proceeds, more of the nervous system comes into play and the body can perform more functions. The baby begins to sit, then to stand, and then to walk.

The brain is divided into many areas, each with a very specialized function (Fig. 5.15). One area deals with movement. An electrical impulse sparked off here will be relayed through a series of nerves and received by a group of muscle cells, which will contract. If this part of the brain is destroyed, or the electrical impulse interrupted anywhere along its path, the message cannot reach the muscle and paralysis results. When a large area of brain involved with movement is destroyed a 'stroke' results. Because the wires cross from one side to the other in their journey from brain to target, an injury on one side of the brain will affect the opposite side of the body.

All computers need to be fed with information in order to function. The brain is already programmed with millions of bits of information at birth. For instance, the facts necessary for the everyday control of breathing and heartbeat are stored deep within its substance. Additional information constantly pours in during the waking hours. This information arrives at the brain via sensory nerves from the eyes, ears, nose, mouth, skin, muscles, and other organs.

If every piece of information was recognized at a conscious level, the brain would immediately break down. Most of it is recorded subconsciously, to be sorted into the patterns of memory during sleep. If this process is interrupted, for instance by waking suddenly, the conscious mind may rapidly try to relate the unconnected pieces of information in the act of sorting, and string them together as a dream or nightmare.

Background (cont.)

The brain recognizes several forms of sensation received from the body through the sensory nerves. They include touch and pressure, temperature, and pain. The degree of sensation varies over and within the body; for instance, the fingertips are very sensitive because they contain many receptors for touch. A map of the surface of the body relating each area with a particular nerve for either touch, temperature, or pain may be drawn. If an area is tested and found faulty the doctor will, by knowing the nerve responsible for the relay of information from the area, be able to determine the site and cause of the fault.

The principal nerves

The nerves that work the muscles of the arm travel from the spine to the armpit (Fig. 5.16). From here, three main nerves, each carrying many fibres, run down the arm between the bones and muscles. These three nerves are the radial, the median, and the ulnar. They carry the impulses needed for the movement of the muscle groups in the arm, elbow, wrist, hand, and fingers, together with the fibres that transmit sensation back to the brain. One of the nerves, the ulnar, runs behind the elbow joint and is easily knocked against the bone, creating a tingling sensation in the fingers—'hitting one's funny-bone'.

The nerves to the legs come from the lower spine. The main nerves are the femoral and the sciatic. It is the sciatic nerve which is often involved in a 'slipped disc' injury, with sciatica as the result. It is the longest nerve in the body and supplies the muscles at the back of the thigh (the hamstrings), the leg below the knee, and the foot.

The femoral nerve is of most importance to the person with haemophilia. It runs down the pelvis and sweeps forward to enter the thigh at the front, at a point halfway along the skin crease between the thigh and trunk. If a finger is placed on this point and pressed gently the pulsation of the femoral artery can be felt. The nerve lies just outside the artery. In its course the femoral nerve lies within a groove between two large muscles, the iliacus and the psoas (pronounced 'so-ass'). These muscles sweep forward like the nerve and end in the thigh, where they are anchored to the femur or thighbone. Because of this close relationship, bleeding affecting these muscles may also affect the nerve (see p. 101). The femoral nerve supplies the muscles at the front of the thigh—the quadriceps, or 'quads'. It is these vital muscles which are weakened so often by bleeds into the back, hip, or knee.

There are two main causes of bleeding inside the skull. The first, and the most important to know about, is injury. Just as a heavy blow to the head will cause bruising of the skin outside the skull it may cause bruising inside the skull. The internal bruising in someone with haemophilia will continue to spread, and as the bleed grows it will start to compress the brain. There are several warning signs that this is

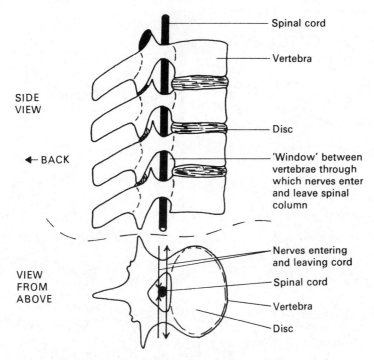

Fig. 5.14. Diagram of part of the spinal column.

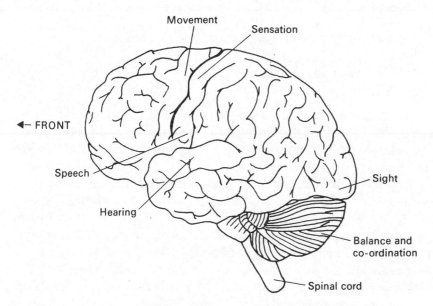

Fig. 5.15. Diagram of the brain from the side.

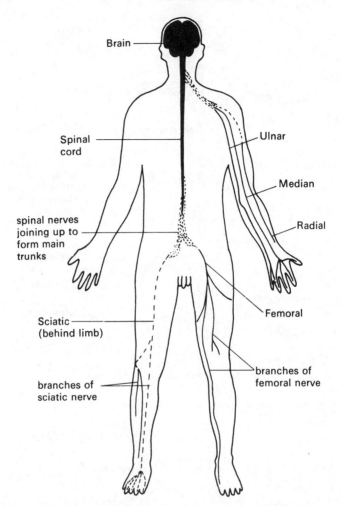

Brain

Spinal
cord

Ulnar

Median

Radial

spinal nerves
joining up to
form main
trunks

Femoral

Sciatic
(behind limb)

branches of
femoral nerve

branches of
sciatic nerve

Fig. 5.16. Diagram of the main nerves to the limbs.

happening, and they may come on very gradually. They include intense
headache, irritability, dislike of bright light, nausea and vomiting, and
drowsiness. There may also be complaints about eyesight and hearing,
and difficulty in moving part of the body.

This type of bleeding commonly arises from blood vessels under one
of the membranes which line the skull, the dura. The bleed is called a
subdural haemorrhage and its treatment is by one of the oldest known
operations. Under anaesthetic, a hole is bored through bone overlying
the bruise and the blood let out. Obviously the sooner this is done the
better, but the surgeon must be sure of the diagnosis and the site of
bleeding before he acts, and special tests will be necessary before the
decision to operate is made.

The second type of bleed is unpredictable and is as rare in people with haemophilia as in anyone else. It starts below another more vascular membrane called the arachnoid, and is called a subarachnoid haemorrhage (Fig. 5.17). Surgery may help these people too but is more difficult than in the case of the subdural. In any event, the first necessity is the urgent treatment of the haemophilia.

Following a bleed inside the skull, the extent of recovery will depend on the amount of brain tissue affected. Like other tissue the brain responds to severe injury by forming scar tissue, and very occasionally this acts as a focus from which abnormal brain waves can originate; the result is epilepsy. Nowadays epilepsy can be controlled extremely well with drugs, and major fits are rare. Another effect may be weakness of part of the body.

The long-term effect of one of these bleeds will not be known for many weeks or months. When part of the brain is destroyed the area around it goes into a state of shock which takes time to wear off. A seemingly hopeless situation at first may resolve into a remarkable recovery with time, and there are many examples of people who have resumed a full and happy life after a stroke.

When bleeding into the head occurs in early life some doctors prescribe prophylactic factor VIII or IX concentrate as a routine, as least throughout childhood and adolescence.

Other clotting defects

These are much rarer than haemophilia and the effects will depend, like haemophilia, on the degree of factor deficiency. Without treatment, prolonged bleeding may follow surgery, childbirth, or major injury.

Factor I (fibrinogen) deficiency: Joint bleeds are uncommon, but there is more likelihood of oozing from small wounds than in haemophilia. Easy bruising will occur.

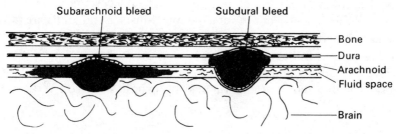

Fig. 5.17. Diagram of a subarachnoid and a subdural haemorrhage.

Factor II (prothrombin) deficiency: Nosebleeds, heavy periods, and easy bruising are features. Joint bleeds are rare.

Factor V deficiency: Easy bruising, nosebleeds, heavy periods, and joint and muscle bleeds are experienced by severely affected people.

Factor VII deficiency: Effects are usually mild but include nosebleeds, heavy periods, and bruising. Joint bleeds are rare.

Factor X deficiency: Bruising, joint bleeds, and heavy periods occur in people with this deficiency.

Factor XI deficiency: Affected people often have Jewish ancestry. Nosebleeds and heavy periods are experienced.

Factor XII deficiency: This deficiency is usually diagnosed only when blood taken for some other reason fails to clot in the normal time. Surprisingly, people with even a severe deficiency have little or no trouble from bleeding.

Factor XIII deficiency: Clots form in the normal time but tend to break down later as they are not stabilized. This leads to a delay in healing after injury or surgery. Females with severe XIII deficiency have frequent miscarriages. Factor XIII deficiency may present early in life with oozing from the umbilical cord. People with severe factor XIII deficiency should be treated with monthly prophylaxis with factor XIII concentrate. This prevents bleeding altogether.

von Willebrand disease and other disorders

In von Willebrand disease both platelet function and factor VIII activity are affected. Because of the platelet defect affected people are more likely to bleed from small cuts and trivial injuries to the linings of nose and mouth than do those with haemophilia, and the bleeding is more likely to be immediate. Easy bruising, heavy periods, and bleeding after childbirth may occur in some people. Muscles and joints are rarely affected. Treatment is with desmopressin (p. 135), or with factor VIII replacement as in haemophilia.

Rare defects in platelet function without factor involvement may cause bleeding like that experienced by people with von Willebrand disease.

6

What to use: therapeutic materials

Although bioengineered factor VIII is now available for treatment in some countries, most of the materials used in the treatment of bleeding disorders in the world are still derived from human blood. This is provided by either voluntary or paid donors and is collected by a transfusion service which may be local or national. In several countries, including the United Kingdom, the donors receive no payment for their generosity. In other countries, most especially in the United States, donors are paid modest amounts for their blood plasma, which is then processed by one of the commercial companies.

In the United Kingdom whole blood is taken at special sessions arranged in localities suitable for the majority of donors in a particular area; these might be in factories, universities, church halls, or recreation centres. Donors already enrolled in the area will often be notified of the session, but new donors are always welcomed without prior arrangement. In order to give blood, a donor must be an adult in good health and not known to carry a communicable disease. Since 1983 donors infected with the human immunodeficiency virus (HIV) have been excluded from donation. Donors positive for hepatitis C in the United Kingdom have been excluded since 1991. The medical questionnaire which a donor must answer each time he or she attends to give blood or plasma is shown in Table 6.1.

Before each donation a simple test ensures that the donor is not anaemic. An attendant checks the donor's card, selects a plastic pack labelled with the right blood groups, and escorts the donor to a couch where he or she is made comfortable. A tourniquet is placed round the upper arm and the elbow crease swabbed with antiseptic. A medical or nursing officer chooses a vein at the elbow and gives a small injection of local anaesthetic into the skin over the vein. A needle is then painlessly

Table 6.1

MEDICAL QUESTIONNAIRE

Please tell the clerk your date of birth

The questions below are designed to protect the health of you, the donor and also to ensure that any patient who might receive your blood does not suffer any ill effects.
If you would like to discuss any item with the doctor, in confidence, please tell the clerk.

A. Since you last gave blood

Have you:
        (a) had any serious illness or operation that has taken you into hospital
        (b) been to a hospital out-patients department
        (c) seen your GP/ family doctor
        (d) had treatment for acne with tablets

B. In the last twelve months

Have you:
        (a) had a tattoo or had ear or body piercing
        (b) had electrolysis or acupuncture
        (c) been in contact with anyone with an infectious or contagious disease

C. Have you ever

        (a) had treatment with hormones to increase your growth
        (b) been at risk of contracting AIDS or the HIV virus
        (c) been abroad outside Western Europe—if so where and when

D. In the last month

Have you:
        (a) taken any medicines or tablets prescribed by your doctor
        (b) taken any medicines or tablets bought without prescription

Declaration by Donor:
I have read these questions and answered them fully to the best of my knowledge.
I confirm that I am not in an AIDS risk group and that I have read the AIDS leaflet.
I agree to my blood donation being tested for the AIDS/HIV virus and other infections.
I understand that these tests are not infallible and I am not donating blood just to obtain a test. If a test is positive I know that I will be informed of the result.

Information on HIV and AIDS exclusion rules will be found on p. 227.

placed in the vein. The needle is attached by a length of plastic tubing to the pack. Blood flows into the pack, where it mixes with a chemical (anticoagulant) which prevents it clotting—the blood and anticoagulant make up a total volume of 520 ml (about a pint).

The needle is removed, gentle pressure put on the site with cotton wool for a while to prevent bruising, and a Band-aid applied. The donor rests for 10 minutes and is given some form of refreshment before leaving.

None of the donation equipment is reused, *so there is no danger whatsoever of a donor becoming infected with HIV or indeed any disease as a result of his or her donation.*

After a session, the labelled packs of blood are taken in a refrigerated van to the central laboratories of the transfusion service. Here they go through a series of stringent tests to exclude infection, including HIV and hepatitis, and to make sure that the blood groups on the labels are correct.

This account describes the collection of whole blood, that is the various cells suspended in their plasma. When refrigerated at 4 °C whole blood is suitable for transfusion for only 5 weeks, even using the most up-to-date technology. After this time the number of red cells will have fallen to levels unacceptable for worthwhile transfusion. Other cellular constituents disappear more rapidly; for instance, active platelets only last 2 or 3 days. This is why donors are called upon to give blood at regular intervals to help maintain fresh stocks.

Many donors supplying the products necessary for the treatment of haemophilia in the developed countries do not give whole blood. They donate plasma only, by a process called plasmapheresis. This is done with special giving sets or machines which allow red cells to be returned to the donor immediately after separation from the plasma. Plasmapheresis donors can donate more often than whole blood donors because they do not become anaemic because of a shortage of red cells. Their plasma is subjected to the same checks as whole blood to exclude disease.

Plasmapheresis used to be carried out using fairly simple plastic giving sets and packs. Obviously the most stringent precautions had to be taken to ensure that the donor got his own red cells back, and in this respect machine plasmapheresis is safer than the bag system. This is because once the donor has been connected to the machine he or she remains connected throughout the procedure, and there is no possibility of receiving someone else's cells. Whole blood from the donor's vein runs into the machine which works in much the same way as a household spin dryer. As it spins, the blood separates into its constituent parts which are then drawn off through convenient portals.

Using a cell separator, red cells, white cells, platelets, or plasma can be collected and the constituents not needed at the time returned immediately to the donor.

In the other method the bag of whole blood has to be cut free from the giving set and taken away from the donor so that the contents can be separated in a centrifuge and the red cells returned. Despite elaborate precautions this method is not foolproof and has, very rarely, resulted in harm to the donor.

Without mass plasmapheresis, especially in Europe and the United States of America, there would have been insufficient blood products to allow the improvements in haemophilia care seen in the past decade.

Component therapy

Nowadays, when whole blood has been donated it is used only for transfusion when a patient is in need of all the constituents, for instance after a very serious injury. Otherwise only the constituent or constituents needed by the patient are replaced. The technique of separating blood and using its parts selectively is called component therapy.

Whole blood contains red cells, white cells, platelets, and plasma (Fig. 6.1). The plasma contains proteins, some of which are necessary for clotting.

If whole blood were given to every patient in need of only one of its components, two things would happen. Firstly, the patient would be overloaded with other components which he or she would not need and which might be harmful in excess. Secondly, there would not be enough blood for everybody. Most blood is, therefore, separated

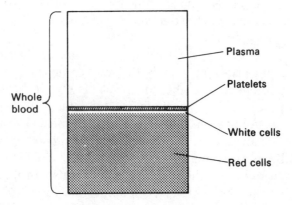

Fig. 6.1. The components of whole blood after separation.

Rotating spindle
with fixed arms

Swivel

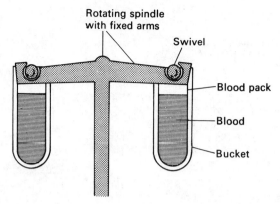

Blood pack

Blood

Bucket

Fig. 6.2. Diagram of centrifuge before separation of plasma from red cells.

into its components and the doctor chooses the right one(s) for the patient's special needs. Separation is performed as soon after donation as possible because some of the constituents only last a short time at 4 °C. Factor VIII is one of these—half of it will have disappeared within 12 hours.

The separation of the plasma from the red cells is simple. A pack of whole blood is placed in a centrifuge (Fig. 6.2) and spun rapidly (Fig. 6.3). Because red cells are heavier than plasma they settle to the bottom of the pack; the plasma can then be drawn off. The packs containing red cells are stored until needed.

The pack with the plasma (Fig. 6.4) may be deep frozen for several months. At low temperatures the natural decline of the coagulation factors is halted. This is fresh frozen plasma and it was the mainstay of haemophilia therapy before more concentrated blood products were introduced.

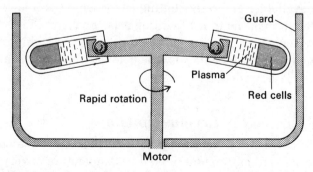

Guard

Plasma

Rapid rotation

Red cells

Motor

Fig. 6.3. Diagram of centrifuge during separation of plasma from red cells.

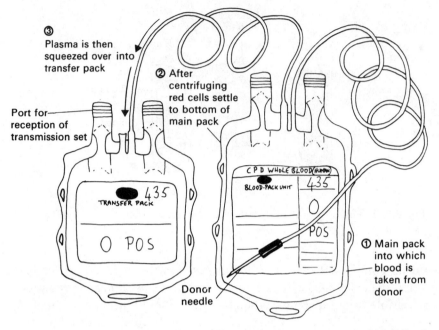

③
Plasma is then
squeezed over into
transfer pack

② After
centrifuging
red cells settle
to bottom of
main pack

Port for
reception of
transmission set

C P D WHOLE BLOOD (HUMAN)

BLOOD-PACK UNIT   435

TRANSFER PACK   435

O POS

O POS

① Main pack
into which
blood is
taken from
donor

Donor
needle

Fig. 6.4. A double blood pack for collection and separation of whole blood.

Fresh frozen plasma (FFP)

FFP is still used in some countries because it is the only product available. However, it is difficult to achieve useful levels of factor VIII, and very difficult to raise the factor IX level sufficiently to stop all but minor bleeds, using FFP. If too much is given in an attempt to force the clotting factor level higher, the circulation soon becomes overloaded and the patient has to be treated for heart failure as well as haemophilia. Provided this is recognized, no lasting harm will come to him, but obviously it is better to use more concentrated products if they are available. The only clotting factor which is unavailable in concentrated form at the time of writing is factor V, but FFP is suitable for factor V replacement. This is because the level at which factor V deficient people stop bleeding is much lower than in haemophilia, hence less FFP is needed.

Cryoprecipitate

If a pack of fresh liquid plasma is plunged into a mixture of alcohol and dry ice (frozen carbon dioxide) the fall in temperature is very sudden and the plasma rapidly freezes solid. When left to thaw slowly at 4 °C,

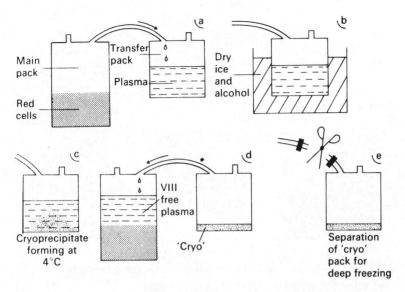

Fig. 6.5. Diagram of the preparation of cryoprecipitate.

a cloudy portion gradually separates out (precipitates). On return to the centrifuge this precipitate settles to the bottom of the pack and most of the remaining plasma can be withdrawn and used for the preparation of other products. The cloudy portion remaining is cryoprecipitate (*cryo* = cold), which is rich in factor VIII and fibrinogen (Fig. 6.5). Cryoprecipitate may be freeze dried. Otherwise it must be kept frozen in a deep freeze in order to maintain its clotting factor activity.

Plasma concentrates

So far the products prepared from single donations have been considered. One donation of whole blood will provide either:

- one pack of red cells and one pack of FFP, *or*
- one pack of red cells, one pack of cryoprecipitate, *and*
- one pack of residual plasma (Fig. 6.6).

The starting point for the preparation of concentrates is pooled plasma usually taken from many thousands of donors. With care to prevent contamination with infection the pool is treated with various chemicals which separate it into a number of fractions, a procedure known as fractionation. It is also subjected to one or more of the methods of viral inactivation (p. 228). The end product is freeze-dried (lyophilized) and, just before use, sterile water must be added to make up a solution for

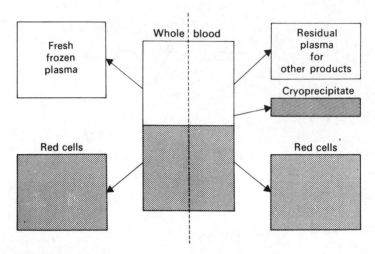

Fig. 6.6. One donation of whole blood provides that shown to the left of the dotted line, or that to the right of the dotted line.

injection. The advantages of the fractionation process are that because the factors are stable in their dry form, the final product does not have to be stored in a deep-freeze, and that a lot of activity is concentrated in a small volume. The disadvantages are that more blood donors are usually required as much of the clotting factor activity (the yield) is lost in processing (Fig. 6.7). Processing is also more expensive and there is, of course, a greater chance of viral contamination of the raw material.

Fractionation is used in the preparation of concentrates of factors

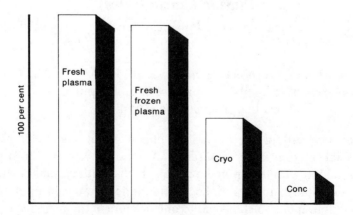

Fig. 6.7. As blood products become more refined the yield of Factor VIII falls (cryo = cryoprecipitate; conc = concentrate).

I, II, VII, VIII, IX, X, XI, and XIII. Many of the products are made in the laboratories of the transfusion services, but because of the enormous number of donors required, the equipment needed, and the cost involved the bulk of the world's supply is made commerically from the plasma of paid donors.

The purity of concentrates

In theory, the treatment of haemophilia is simple. The only difference between someone with haemophilia and someone without it, is the lack of an active clotting factor. Therefore, all that is needed is replacement of that factor.

Blood plasma contains many different proteins in addition to the clotting factors. Someone treated with plasma, or with cryoprecipitate, receives far more than factor VIII. Most of the protein transfused is simply discarded by the body, without any harmful effect. There is no evidence that people suffer in any way in the short term from the effects of these proteins after multiple plasma transfusions.

However, someone with severe haemophilia A or B who needs regular treatment with factor VIII or IX over a lifetime is being exposed to far more of this extraneous protein than anyone else in the history of medicine. By the time he is 70-years-old he will have received at least 3000 transfusions each extracted from thousands of donations of plasma. If he has been on prophylaxis or needed major surgery his exposure will be all the greater. So there is a valid reason for arguing that he should be treated only with products containing little or nothing more than the relevant clotting factor.

The differences between the products used in treatment are expressed in terms of purity. A concentrate containing only factor VIII would be of high purity, whilst plasma would be least pure. In between are products with variable amounts of 'gunge' that are said to be of intermediate purity.

There is, however, a major problem in all this. By itself factor VIII is a very fragile and unstable beast. If it is to last in the vial and be useful for haemophilia therapy it has to be stabilized. Various ways of doing this are used by the manufacturers of concentrate, but one of the best is to add albumin, a major protein made from human plasma, to the product. This, of course, defeats the original purpose of the exercise. We are still transfusing a lot of worthless protein in addition to the factor VIII. Paradoxically even the first generation of synthetic, genetically engineered recombinant products are stabilized in this way.

A further difficulty, more important in the past before the introduction of effective ways of removing viruses from concentrates, concerned yield. The yield is the amount of useful factor VIII or IX in the end product, compared with that in the source plasma. In general the more steps required to manufacture a concentrate, the lower the yield. For instance, by the time plasma has been collected and deep frozen 5 per cent of the active factor VIII has been lost, and making cryoprecipitate loses a further 50 per cent. One result is the need for more donations of plasma. And more donations means more donors and more exposure to the risk of infection.

Advances in manufacture currently in the pipeline address these problems. A new, so-called second generation recombinant factor VIII is being made without the need for an albumin stabilizer. Ways of ensuring the elimination of all viruses, whether or not they have fatty coats, are being explored. Improvements in yield are already being achieved.

But what does all this mean in practical terms to the person with haemophilia? Is he likely to be better off if treated only with the purest product available? And are there significant differences between the product of one manufacturer and another?

At the time of writing the truth is that we do not really know; the jury is out. Those strongly in favour of using only products of the highest purity argue that there is not time to wait for a verdict. They say that by the time a verdict comes, if it ever does, it will be too late for people with haemophilia who will in the interim have received less pure products. Those strongly in favour of waiting point to the lack of scientific evidence of any harmful effect in people exposed to less than pure products for many years. To date the only adverse effects reported have been in laboratory experiments, and not in real life.

The only exceptions to this are in people with HIV infection, and people with haemophilia B. There is sufficient evidence to be able to justify offering high purity concentrates to HIV positive people with haemophilia. The evidence we have shows that these concentrates have less of a detrimental effect on their immune systems. Purer products for haemophilia B are justified because they are probably less likely to cause thrombosis, especially after major surgery, than the earlier products which contained factors II and X, in addition to factor IX.

There are other points in this complex argument. There is evidence that people treated with high-purity concentrates, whether of human or synthetic origin, may be more likely to develop clotting factor inhibitors. There are reports of inhibitors appearing for the first time after switching to a particular concentrate despite years of previous, successful therapy with products made in a different way. Most of the high-purity products, including recombinant factor VIII, are far more expensive than earlier concentrates. The more human plasma that is used to manufacture products of the highest possible purity for the richer nations, the less final product there will be available for developing countries.

In summary, present hard evidence suggests that:

- HIV-infected people with haemophilia should be offered treatment with 'high-purity' concentrate.
- People with haemophilia B be treated with a high-purity factor IX concentrate, especially after surgery.
- Whilst the jury is out, consideration be given to starting all newly diagnosed children on a high-purity concentrate.
- That other people not on high-purity concentrates have nothing to fear.

In addition, there are valid reasons for using the newer products for the management of major surgery or complications because they are easier to make up and give quickly.

Finally, there is a continuing discussion about the value of volunteer versus paid donor products. The discussion centres on two particular points:

- Plasma from volunteer donors, who are not paid for their donations, is less likely to contain infectious organisms.
- It is intrinsically wrong to pay for a human resource.

Whilst I have no doubt that the first point is true, it can now be argued that the difference in infection rates between volunteer and paid donors is immaterial. Provided manufacturers test each individual donation, and provided that all infection is eliminated during manufacture, it should not matter where the source plasma comes from.

In my opinion no one will ever win the argument centred on the second point. Whilst altruism is obviously something to aspire to, in the real world there is simply not enough of it to go round! People with haemophilia must have access to safe, high-quality products in sufficient quantities to protect them from the consequences of bleeding. They must also have the right to choose the best product available for a particular problem at a particular time. History has shown that a market restricted to one product because of total reliance on the volunteer principle fails to meet these demands.

This field changes so quickly that any comment is bound to be outdated not long after this book is published. The best advice is to talk to the staff at your haemophilia centre. They will be able to cut through the gloss of advertising and politics and give you current, objective information on which you can make up your own mind about the treatment you want.

Recombinant factor VIII

Since the last edition of this book was published in 1990, a number of manufacturers have successfully used genetic engineering to make and market factor VIII. The resultant recombinant products behave just like the factor VIII derived from human plasma. Recombinant factor VII and IX are in preparation. To understand how these products are made it is necessary to turn to our knowledge of the gene.

Genes are lists of information inherited from our parents that tell our cells how to function. Like beads on a necklace they are strung along structures called chromosomes. Each cell in the body contains a full complement of genes, but not all are active. Which genes are switched on or switched off determines the activity and thus the function of a particular cell. For instance, under their inbuilt genetic control primitive stem cells deep in the bone marrow begin to differentiate into the various mature constituents of blood, whilst in the liver some cells are busy making and secreting factors VIII and IX. It is this slave-like commitment to a single task that led to the idea of manipulating sets of genes so that they would instruct cells to build biologically active products to order. How is this done?

Once the correct sequence of amino acids in a protein is known it is possible to read back to the genetic code that created it and then to use this code to produce new supplies of the protein. To do this a manufacturer is required, and the best manufacturer is one with a rapid turnover. Bacteria provide the simplest answer because they reproduce so rapidly, and are relatively easy to grow in artificial cultures. If a bacterium can be 'persuaded' to accept

Table 6.2 Summary of human products available for coagulation factor replacement. Those made from pooled donations are lyophilized (freeze-dried) and reconstituted with sterile water. This table does not show the presence of other ingredients in the products.

Single donations	Fresh whole blood	All factors
	Fresh plasma	All factors
	Fresh frozen plasma	All factors
	Cryoprecipitate	Factor VIII and fibrinogen
Pooled donations	Fibrinogen	Factor I
	Factor IX concentrates	Factor IX
	Factor II, IX, X concentrate (freeze-dried)	Factors II, IX, X
	Factor II, VII, IX, X concentrate	Factors II, VII, IX, X
	Factor VIII concentrate	Factor VIII
	Factor XIII concentrate	Factor XIII

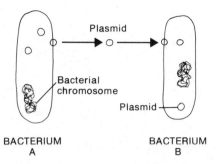

Fig. 6.8. Combining two sets of DNA instructions by introducing a plasmid (bacterial genetic material) from one bacterium to another, this results in recombinant DNA.

the genetic blueprint for a protein, it and its progeny will make it.

In order to introduce the new blueprint into the bacterium its own genetic structure must be manipulated. In some bacteria genetic material lies outside the nucleus in the form of rings called plasmids. Using a chemical it is easy to introduce a plasmid into another bacterium which accepts it as itself (Fig. 6.8). This is the technique of combining two sets of DNA instructions, one native to the cell and the other foreign to it. The resultant product is called recombinant DNA (see below).

Using specific enzymes, which chop up lengths of protein, a plasmid is cut open (Fig. 6.9).

Into the gap created is placed the length of genetic code known to be the blueprint for what we want to make (Fig. 6.10). This length sticks to the cut ends of the plasmid, restoring the circle and recombining the DNA.

The plasmid is now introduced into a bacterium, commonly a strain called *Escherichia coli* which lives happily in laboratory culture as well as in the human bowel.

Whenever the bacterium reproduces itself, so does the plasmid. And the recombinant DNA instructions it carries make the new protein (Fig. 6.11).

In the case of factor VIII, the molecule is too big for bacteria to manage and larger mammalian cells grown (in special cell culture medium) from Chinese hamster ovaries or kidneys are used instead. However, the principle is the same as that described above. If the process is scaled up to a commercial level human factor VIII should be available in huge quantities.

Unfortunately, like most things in life, the story is not as easy as

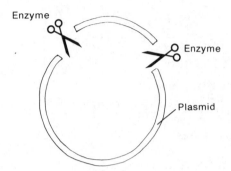

Fig. 6.9. Preparing a plasmid for inserting the new instructions (see text for further details).

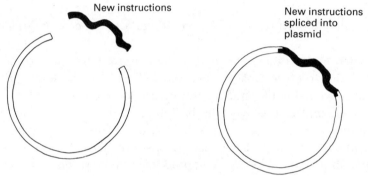

Fig. 6.10. Inserting the new instructions into a plasmid (see text for further details).

that. Not only have all the genetic instructions to be absolutely right, and the mammalian cells which make the factor VIII kept alive and well, but stringent safety precautions have to be taken in order to clean away all impurities from the finished product. In the methods presently available these impurities include material from both the hamster cells used to make the product, and from mouse antibodies and bovine serum used during manufacture. And as if that were not enough, because the clotting part of the factor VIII molecule is so fragile, it has to be stabilized in the final bottle by albumin from human donors in two of the three recombinant products now available. However, after 5 years of clinical trial no ill effects, other than a possible increase in the incidence of clotting factor inhibitors (p. 205), have been reported.

The production line for genetically engineered factor VIII is shown in Fig. 6.12. A master bank of cells containing the original recombinant instructions is kept, and used as a start point at regular intervals. In this way no manufactured product can get too far away from the parent

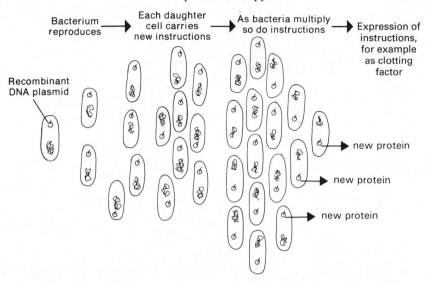

♂ = new protein instructions

Fig. 6.11. Diagram to show new protein production as a result of recombinant DNA instructions.

recombinant material, and changes in genetic structure (mutations from the parent cell line) are less likely to occur.

Animal products

In some countries blood from pigs and cows has been used as the starting-point for commercial factor VIII concentrate. The products, known as porcine anti-haemophilic globulin (AHG) and bovine AHG, respectively, are potent freeze-dried powders. Only porcine factor VIII (Hyate:C) is now made by Porton Products Limited in the United Kingdom.

Hyate:C is an interesting product with great potential. The original pig VIII concentrate caused a fall in platelets in the recipient's blood. The new product does not. The original concentrate caused resistance in the recipient about a week after a course was started. The new product does not do this on every occasion. If the cause of resistance, which results in a failure to raise the level of active factor VIII, can be identified and overcome, the porcine material will be a more viable and potentially safer product for the treatment of haemophilia A. At present it holds a well-deserved place in the management of people with factor VIII inhibitors (see Chapter 10). It should also be considered as an alternative to human product for moderately or mildly affected people with haemophilia A who need only infrequent therapy, and who

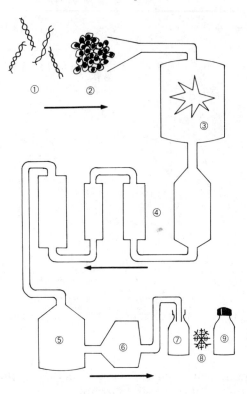

Fig. 6.12. The manufacture of recombinant factor VIII. (1) DNA instructions for factor VIII are inserted into the cells (2) which make factor VIII continuously in a vat containing oxygen, proteins, and carbohydrates (3). The recombinant factor VIII is harvested from this vat (4), and impurities are removed in a series of filters and columns. Stabilizing solutions are then added (5) before final filtration (6) and vial refilling (7). Freeze drying (8) and capping and labelling individual vials complete the process (9).

Instructions for the clotting part of the factor VIII gene, synthesized from knowledge of the sequences of human DNA, is inserted into hamster cells, either grown from kidney or ovary. Thereafter these cells will make and express clotting factor VIII in addition to their usual products. The factor VIII now needs to be removed from these other products, and from the cells themselves. This purification is done in a number of ways, including heating, filtration, and by a method using specific, or monoclonal, antibodies which bind exclusively to the factor VIII. These antibodies attract pure factor VIII from the mixture, rather like a magnet attracting iron filings. A further step frees the factor VIII from the antibodies.

After a final filtration a stabilizing protein (at present human albumin) is added before the bottles are filled and the product freeze-dried and labelled.

do not get an adequate response to desmopressin (see below). It should be noted that the present product, while free from human viruses, does not undergo antiviral treatment during manufacture. This is likely to change in the near future.

Jehovah's Witnesses

While members of this faith have deep convictions against the transfusion of whole blood and some of its constituents, they do not prohibit the use of anti-haemophilic preparations. Alternatively, the recombinant product is available for people with haemophilia A. However, only the second generation product developed by Pharmacia in Sweden is free from human albumin.

References to blood products and Jehovah's Witnesses may be found in the *Journal of the American Medical Association* (1981), vol. 246, pp. 2471–2.

Desmopressin (deamino-D-arginine vasopressin; DDAVP)

Vasopressin is one of the hormones produced by the pituitary gland which lies at the base of the brain. Its main action is to help conserve water in the body by acting on the kidneys, hence its other name, antidiuretic hormone (diuresis = promotion of secretion of urine). Desmopressin is a synthetic derivative of the natural hormone, developed originally to help people with disease or injury of the pituitary gland. It may be given by injection into a vein or under the skin, or by nasal spray.

Coincidentally, desmopressin raises the level of factor VIII in the bloodstream. This rise is sometimes sufficient to stop bleeds and to allow even major surgery to be performed. Desmopressin is only effective when the person receiving it is already capable of producing some normal factor VIII. Thus it works in mild haemophilia and for most people with von Willebrand disease, but not in severe haemophilia. It has no effect on the other clotting factors.

Desmopressin does not carry the side-effects associated with blood products like hepatitis or HIV and has, therefore, gained a very useful role in haemophilia management. Because it also enhances fibrinolysis an antifibrinolytic drug is often given at the same time. The main side-effect of desmopressin is water retention and people treated with it should restrict their intake of fluids for a while. Desmopressin should be used with caution in infants, and in patients with heart or kidney disease.

Whilst desmopressin is usually of value in von Willebrand disease, it is contraindicated in some forms of this disorder because it can have an adverse effect on the platelets (p. 87).

Ways of giving desmopressin

Desmopressin may be given by intravenous (IV) or subcutaneous (SC) injection, or via the nostrils (intra nasal, or IN, administration). The drug is made and marketed under a variety of names, including DDAVP, Desmospray and Octostim, by Ferring. It must be stored in a refrigerator at 4 °C–8 °C.

- Intravenous injection is given slowly by a drip. It is presently the usual way of giving the drug in hospital but, with newer preparations, this practice is changing. Intravenous injection will probably still be used before surgery or in order to treat major injuries because it ensures that the dose has entered the body in sufficient quantity to raise the factor VIII level.
- Subcutaneous injections are given just below the skin, usually in the abdomen or the front of the upper thighs. They are often more convenient than IV injections and are suitable for home therapy or holiday use by some people.
- Intranasal desmopressin is set to become the most popular form of the drug. The membrane which lines the inside of the nose is very rich in blood vessels. Desmopressin crosses this membrane easily and enters the bloodstream as effectively as if it had been injected. However, a very highly concentrated solution of desmopressin is needed to do this. This preparation is now licensed for use in some countries. It is expected that improvements in manufacture and supply will soon extend its use generally.
- IN desmopressin can be easily self administered by the person with mild haemophilia or von Willebrand disease. A couple of drops in each nostril are sufficient to cause a rise in factor VIII. IN desmopressin is therefore ideal for home therapy.

Local haemostatic agents

These are substances which are prepared for direct application to a bleeding area. They work either by causing blood vessel contraction, by inducing rapid clotting at the site of injury, or by sealing the wound artificially. Some are on direct sale to the public in the form of sticks or foams to be used for staunching troublesome bleeding after wet razor nicks. Available to the doctor are adrenaline, which contracts the vessels, and topical thrombin, which converts fibrinogen to fibrin. In some countries, Russell's viper venom, which short-circuits the clotting mechanism, may still be available. As the name suggests this product comes from Russell's viper, an Indian snake.

Unfortunately, these products are useless in the treatment of all but the most minor injuries. Their action is very short-lived and they are quickly washed away by the blood. Even if an initial cessation of bleeding is achieved, blood usually breaks through the flimsy barrier after a minute or two. If they are applied on cotton material bleeding starts as soon as this is removed; materials like Oxycel, Sterispon, or Surgicell, which dissolve away gradually, are better. Local haemostats are occasionally useful in the treatment of nosebleeds (see also p. 138), bleeding from tooth sockets, and minor abrasions of the skin. They cannot be injected as they would cause widespread clotting throughout the body.

Another sort of local treatment has recently been developed and marketed under a variety of names. Fibrin glue relies for its action on the final stage of the normal clotting process, and is more efficient than topical thrombin on its own. It has uses in surgery quite apart from haemophilia. A new technique which allows thrombin to mix with other clotting factors directly at the site of bleeding is under trial in Israel. This 'superglue' may not only help to reduce the amount of factor replacement needed for surgery, but also provide a new way of helping people with haemophilic pseudotumours, in whom it might be used for sealing cavities after removal of the pseudotumour.

Antifibrinolytic drugs

These work by preventing the natural breakdown of blood clots which have already formed. They do this by blocking the mechanism of fibrinolysis (p. 54).

The two most widely-used antifibrinolytic drugs are EACA, which stands for epsilon amino caproic acid, and AMCA, which stands for amino methylcyclohexane carboxylic acid. EACA is marketed under the name Epsikapron and Amicar, and AMCA, also known as tranexamic acid, is marketed as Cyklokapron.

Because the action of these drugs (Fig. 6.13) depends on there being a clot at the site of injury in the first place, they are of no value in the initial treatment of patients with severe clotting factor deficiencies. Their greatest value is after dental extractions once a clot has formed following clotting factor transfusion. They are also useful in the management of open wounds, whether surgical or accidental, and in the treatment of heavy and prolonged periods. Some people with haemophilia or von Willebrand disease find that a couple of

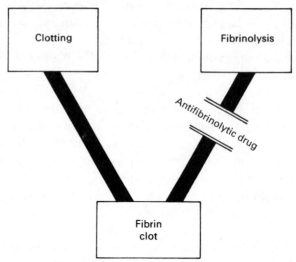

Fig. 6.13. Site of action of antifibrinolytic drugs which stop the breakdown of blood clots.

drops of Cyklokapron solution placed in the nostril helps to control nosebleeds (p. 92). The antifibrinolytics do not seem to have much effect on closed injuries like haemarthroses.

The drugs may be dangerous under certain circumstances and must never be taken without medical guidance if there is blood in the urine or if there is any evidence of kidney damage. In these cases, clots may form in the collecting systems of the kidneys and urine may be dammed back. Subsequent passage of the clots is extremely painful.

Antifibrinolytics also increase the effectiveness of some factor IX concentrates to the point where inadvertent clotting occurs. Apart from their use in the nose to control epistaxes, they should not, therefore, be used at the same time as factor IX concentrates.

7

How to treat: home therapy, prophylaxis, venepuncture

Indications for treatment

Treatment should be given as soon as possible for the following:

(1) bleeding into a joint;
(2) bleeding into a muscle, especially in the arm or leg;
(3) injury to the neck, mouth, tongue, face, or eye;
(4) severe knocks to the head and unusual headache;
(5) heavy or persistent bleeding from any site;
(6) severe pain or swelling in any site;
(7) all open wounds requiring stitches;
(8) following any accident that may result in a bleed.

True emergencies are very rare in the bleeding disorders. Because bleeding is persistent rather than profuse, there is usually a delay of two or three hours from the onset of a bleed to its full development. This gives time for either getting the person with haemophilia to hospital or giving home therapy. First aid measures apply in exactly the same way as in anyone without haemophilia.

In general all the complications listed above, with the exceptions of 1, 2, and 8, should be seen by a doctor with haemophilia experience. Home therapy is suitable for joint and muscle bleeds, small injuries and cuts not needing stitches, and nosebleeds. Open bleeds in people with haemophilia A should usually be treated with an antifibrinolytic as well. Haemophilia B patients are cautioned about the use of these drugs in conjunction with factor IX concentrates (see p. 138).

Home therapy should never be the only treatment of patients with injuries to the head, neck, or eye, or with severe internal pain, especially if this is in the abdomen.

The initial treatment of painless haematuria (blood in the urine) can be carried out at home. Antifibrinolytics must be used with great care in haematuria because they may lead to clotting in the urinary tract with blockage of normal urine flow and acute pain.

The earlier the treatment of a bleed is started the better.

Early treatment prevents the effects of pressure on surrounding tissues. It prevents the disruption of tissues. Less blood will find its way into joint spaces. There will be less pain. Rapid resolution of bleeds with a quick return to activity and an absence of long-term complications will result. Less blood product will be needed to control the bleed.

> An analogy I use with children is to get them to imagine that a bleed is a fire. If the fire is small, say a camp fire, pouring a bucket of water over it will put it out. If the fire is allowed to spread much more water will be needed to put it out. If it is allowed to reach the trees of the forest firemen and engines, and even aeroplanes, drenching the flames with water will be needed. And there will be damage for a long time. The same is true of a bleed. Bleeds treated very early, when they are small, stop. Bleeds left to spread are not stopped easily, need much more treatment, and may result in long-term damage.

People with haemophilia learn by experience to tell when they are bleeding before any outward signs of a bleed are apparent. I borrowed the word 'aura' from those people who suffer from epilepsy and who know in advance when they are going to have a fit, to describe the premonition of a bleed. In haemophilia an aura is the 'funny', 'fizzy', or 'trickling' feeling that announces the start of a bleed. The knowledge that an aura precedes a bleed begins to form in childhood. But a child who is frightened by the thought of treatment, or of punishment from his parents for 'causing' a bleed (which might well be spontaneous) will sometimes hide it until he is forced to seek help because of pain or swelling. Joint and muscle bleeds later cause both pain and swelling and also limitation of movement. As explained in Chapter 5 the limb will take up its position of maximum comfort and should not be moved from this until the acute bleed has started to settle.

Very young children may show the first evidence of a bleed by restricting movement of the affected limb, and the first signs of recovery by moving it again. This is the usual reaction of any child, or indeed animal, who senses that something is wrong with an arm or leg. Injury, infection or, in the case of haemophilia, bleeding quickly result in immobility of the affected area. Usually there is nothing to see, and very young children may not be able to describe what is wrong. They may or may not cry, and may or may not want to continue with play. They may lose their appetite, look pale, and complain of tenderness in a particular area on pressure. Or they may not! The best clue is the immobility— the refusal to move a limb by a child holding it in a position of comfort. Taking off or putting on clothes may lead to unexpected tears. A good way of testing any child in order to find out the site of pain is to gently examine him or her during sleep. If he really hurts he will soon let you know!

Once a major bleed is known to have started it is useless to try to forget about it in the hope that 'it will just go away'. It will only stop with the right treatment or with pressure. Resting for days or weeks in bed will not only interfere with normal activities, education, and work, but will cause muscle weakness and the likelihood of further bleeding. If in doubt the advice of a doctor at the haemophilia centre should be sought.

The management of specific bleeds
Haemarthroses

Joint bleeds (haemarthroses) are very variable. They may grow very gradually without pain—the joint becomes full but not tense, and without treatment the swelling takes a long time to resolve. Movement may be restricted only partially, if at all. They may start suddenly, often during the night, the joint filling rapidly with blood and becoming tense, hot, and painful. They may recur frequently with so short an interval between bleeds that there is no time to build up muscle power again.

Whichever type of joint bleed is experienced, the most important treatment is the immediate replacement of the missing factor. If a small dose is given as soon as the person experiences the aura (feels he is bleeding) the haemorrhage will usually cease with no after-effects. The longer factor replacement is delayed the more blood will enter the joint and the more treatment will be required.

Before going for treatment, or if there is going to be a delay in giving home therapy (and if it does not hurt too much) a crêpe bandage should be firmly wrapped round the affected joint. This will help the pressure of the tissues to contain the bleed. Some doctors will aspirate a joint just before giving the factor. I do not recommend this treatment because damage to the lining of the joint or infection may result, and early treatment is an effective alternative. However, when aspiration is done there is usually an immediate relief of pain because the joint capsule is no longer stretched. Adequate factor replacement must follow aspiration, which may in itself cause additional bleeding.

The question of splinting or early exercise is a matter for the doctor and patient. Short-term splinting in a well-padded plaster or plastic back-slab can sometimes ease pain considerably by holding the joint in a position of comfort and allowing the muscles, which have been guarding the joint from pain caused by movement, to relax. When appropriate, replacement therapy should precede splinting. Recovery is often helped by continuing to wear the splint at night for a while. If bleeds have started during the night a night-splint will often prevent a run of haemarthroses without interfering with muscle action during the day.

Many adults dislike splints and prefer to rest the limb on a pillow until the bleed has started to settle. Some like to start exercise immediately after treatment. These are matters of personal preference, but in general the sooner exercise is started the better, even if one limb is splinted. The worst thing to do is to allow the rest of the body to become flabby whilst waiting for one part of it to improve. That is asking for trouble with bleeds in other joints.

Bleeds are more likely to affect the hips, the knees, and the ankles when the person is fat. The joints must bear more weight and obesity makes people more prone to falls and other injuries. Keep slim!

Haemarthroses do not usually require very much factor replacement if treatment is started early enough. In haemophilia A a rise of factor VIII to around 25 per cent is sufficient to control most bleeding and one injection may be sufficient. Sometimes a second dose is necessary if pain is still present or if the joint is tender, a sign which seems to correlate well with active bleeding. In one trial in Newcastle we found that 80 per cent of acute bleeds stopped with a single treatment given soon after recognition of a bleed. The remaining 20 per cent needed additional therapy, but all stopped with four injections of clotting factor.

Muscle bleeds

These nearly always follow a recognized injury and, if factor replacement is given as soon as possible after a heavy blow or sprain, little trouble will follow. When a muscle bleed is left to progress not only will muscle fibres be killed, but the joints worked by the muscle will be affected. Muscle bleeds are the major cause of visible crippling in people with haemophilia because they pull joints out of their optimum working positions. Chronically swollen joints may be hidden by clothing, but the limp of a man who cannot put his heel to the ground because of untreated calf bleeding is visible to all.

If the bleed is large and in an area where the surrounding tissues do not exert much pressure, a collection of blood will pool in the muscle and take weeks to resolve. Even then cysts may form as a permanent aftermath of the bleed and, rarely, a pseudotumour may result.

Like haemarthroses, muscle bleeds are easy to treat with factor infusions if these are given early enough. One infusion is often sufficient and the person need not be admitted to hospital unless the bleed affects the psoas/iliacus muscles, when frequent infusions will be necessary. When large muscle groups like the quads are involved a secondary bleed 7–10 days after the first bleed is not uncommon. It is worth considering further treatments on days 7 and 10 to prevent this happening. Early exercise is very important after a muscle bleed in order to prevent contracture. I have found that treatment with the Curapuls (a pulsed short-wave diathermy machine) by the physiotherapist can be of great benefit in some muscle bleeds.

Neck, mouth, tongue, face, or eye

Early treatment is necessary in these situations because bruising may be extensive in the loose tissues of the head and neck and may start to interfere with breathing. If the bleed has started from a sore throat or other inflamed area, the infection should be treated with an antibiotic. Factor replacement is required and rather higher doses than those used for joint and muscle bleeds will be given. The patient may be kept in hospital for observation for a few days as a precaution against secondary haemorrhage. As well as factor therapy and antibiotics other drugs may be used in cases of severe swelling. Steroids, which reduce inflammation, and diuretics (drugs which reduce the volume of fluid in the tissues by increasing the output of urine) may be helpful in the early stages of treatment.

Tongue bleeds in children often follow small bites, and bleeding will be very persistent without adequate factor replacement. Local measures to try to seal off the wound with haemostatic agents are useless and stitching will make matters worse. Every time a clot forms it will be knocked off again by sucking or movement against the teeth. One injection of a moderate dose of the appropriate factor, together with antifibrinolytic treatment (p. 137), will often solve the problem without need for admission. Following wounds to the tongue and inside of the mouth, and in the days after dental extractions, hard foods like toast, biscuits, and potato crisps and very sticky substances like toffee or nougat, should be avoided. Some people find that sucking an ice cube helps stop minor bleeds in the mouth.

Injuries to the eye must be treated very seriously—as they are in any person. Factor replacement will be required at the slightest suggestion of bleeding behind or around the eye. Bleeds into the eye, visible as small collections of red blood to one side of the pupil, are fairly common and usually settle down quickly after one treatment. Bleeds which interfere with vision are rare but do require urgent, expert attention.

Head injuries

The treatment of these has been referred to in the chapter on bleeding episodes. Urgent hospital admission and factor replacement are necessary if any of the signs of internal bleeding are present. Because secondary bleeding may occur, the duration of the treatment is likely to be prolonged. Antifibrinolytic treatment is usually advocated as well as factor replacement.

Heavy or persistent bleeding from any site

The three common bleeds in this category are those from the nose, from the urinary tract, and from the womb.

Nosebleeds

These are often associated with colds or nose-picking and come from a nest of small vessels in an area called Little's area on the side of the central septum just inside the nostril. If a yellow-green discharge is present antibiotic therapy may be indicated, but it is valueless in the common cold with clear discharge as this is caused by a virus and not by an antibiotic-sensitive bacterium.

Nosebleeds will often stop when firm pressure is applied to the affected nostril for 5–10 minutes with a finger. The patient should be sitting up, as in the lying position there will be a greater pressure of blood in the veins. If fainting occurs during a nosebleed he should, of course, be laid down. If bleeding persists despite two or three tries with pressure, factor replacement will usually be required.

Persistent nosebleeds are often treated with packing, in which a length of tape soaked in Vaseline is put up the nostril. This is unlikely to be effective in people with bleeding disorders because as soon as it is disturbed bleeding will restart, but if bleeding is very heavy it is worth a try. Similarly cautery, in which a hot wire seals off the vessels, or laser therapy is often only of temporary benefit. Sometimes the local administration of antifibrinolytic drops to the area from a syringe, followed by pressure, works in haemophilic nosebleeds, and sometimes strong salty preparations may bring relief. When these measures do not work, factor replacement with a few days' antifibrinolytic treatment is the answer.

Bleeding in the urine

Haematuria is a frequent complication in some people with haemophilia. The amount of blood lost is small but it stains the urine bright red. This is the one occasion in the bleeding disorders where a bleed may be left for a couple of days without treatment to see if it resolves naturally. The only proviso to this advice is that early treatment should be sought if the bleeding is associated with pain or if the patient feels unwell. Persistent and frequent bloody water sometimes implies an underlying infection or other abnormality apart from the bleeding disorder and may need investigation. If bleeding has not abated after 48 hours factor replacement will be necessary. A small dose is usually sufficient, but occasionally steroid therapy for a few days is required to stop the bleed. Antifibrinolytic therapy may cause clots to form in the kidneys and is usually contraindicated.

Bleeding from the womb

Heavy and persistent periods may trouble women with a bleeding disorder like von Willebrand disease or factor V deficiency, and are occasionally a nuisance in haemophilia carriers. A hormone pill like that used for contraception, or the monthly use of an antifibrinolytic drug for a few days usually puts matters right (see Chapter 15).

Severe pain or swelling in any site

A bleeding disorder does not prevent other diseases, and the person with haemophilia is just as likely to have appendicitis or an abscess as anyone else. People should always seek early treatment and, if away from home and in a strange hospital, make sure that a senior doctor and the senior nurse in charge understand that they have a bleeding disorder. This is where a special card like that issued by the United Kingdom Department of Health, or the World Federation of Hemophilia, or an identity disc (p. 35) is of great value. Bleeds may mimic other disorders and other disorders may mimic bleeds, so it is vital that patients and relatives are persistent in making sure that medical and nursing staff know of the underlying disorder. If in doubt they should contact the nearest haemophilia centre, or their own centre, for help immediately.

Open wounds

Many lacerations and deep cuts can nowadays be closed with special sticky tape (Steristrips or butterfly dressings) instead of stitches, but factor replacement is usually still necessary as well unless the wound is small. Non-stick dressings should be used to cover open wounds. If ordinary gauze or cotton wool is used the healing wound will be disturbed when the dressing is removed and bleeding may restart. For obvious reasons careful management is particularly important in open wounds to the face. In large hospitals a surgeon skilled in plastic surgery will usually deal with these in children, but whichever surgeon is involved, the patient will need the protection of replacement therapy to reduce swelling and allow a firm join. Continued bleeding below the skin will disrupt stitches and leave a scar.

One of the dangers of open wounds is contamination with the germs of tetanus (lockjaw). This is a terrible disease, needing weeks of intensive medical and nursing care, and the death rate is very high. Because surgery is needed to create a temporary artificial airway (a tracheotomy) in patients with tetanus, prevention of the disease by immunization is of particular importance in people with bleeding disorders. Tetanus can be prevented by immunization with tetanus toxoid; three small volume intramuscular injections are required with booster doses every 8 years. Immunization is very safe and the small injections required will cause no trouble if firm pressure is applied to the injection site in the upper arm for 5 minutes after the procedure. If swelling occurs, and this is very rare, a dose of the appropriate factor will be required.

Following any accident that may result in a bleed

If in doubt, treat!

Other bleeds

Bleeding from the bowel always requires hospital treatment. Peptic ulcers are usually treated by medical means involving diet, drugs which heal the ulcer directly or reduce acidity and spasm, and the avoidance of alcohol and smoking. Factor replacement and antifibrinolytic therapy will be necessary initially and, of course, if an operation is thought desirable.

Cimetidine (Tagamet) and ranitidine (Zantac) and two newer drugs called famotidine and nizatidine are commonly prescribed ulcer-healing drugs; they work by shutting down acid production in the stomach.

Other medicines which may be recommended are carbenoxolone (Biogastrone, Duogastrone) and a form of bismuth (De-Nol). Misoprostol is a drug thought to be especially effective in the treatment of ulcers caused by non-steroidal anti-inflammatory agents (NSAIAs, see p. 274). All these preparations are safe in people with bleeding disorders. However, like all drugs, they do have side-effects and most should be taken only for short periods of time at full dosage (usually 6 weeks), under medical supervision. It is probably unwise to stop treatment abruptly; rather the dosage should be reduced gradually at the end of a course. Sometimes a small dose at night may be used for maintenance of the healing effect.

Whenever a diagnosis of peptic ulcer is suspected certain investigations are indicated, especially in older people. They include endoscopy, in which a flexible tube is passed down the gullet under clotting factor cover. An ulcer can be seen through the endoscope and, if necessary, a biopsy taken from it under clotting factor cover. Recent evidence has linked the organism *Helicobacter pylori* with ulcers and antibiotic treatment may be recommended. There is no contraindication to this in someone with haemophilia.

Home therapy

Many families with haemophilia have now learnt the common-sense way of managing the disorder—home therapy. The intravenous injections of the factor concentrates necessary are soon mastered providing good training is given and the parents and/or patient are relaxed and comfortable with the techniques involved. Depending on the programme of therapy prescribed by the hospital specialist these injections might be given only at the beginning of fresh bleeds, or at regular intervals in an attempt to prevent bleeds, a procedure known as prophylaxis or maintenance therapy (p. 163).

Home therapy is suitable for most severely affected patients and for some people with moderate or mild disorders. People with high-titre inhibitors and some very young children with difficult veins may have to be excluded for a while. Some people are not suited to giving their own injections for temperamental or other reasons. The decision to include a patient in a programme must be made by the doctors who know him in consultation with the family.

Patients with haemophilia A or B (Christmas disease) for whom factor VIII and IX concentrates are available are those for whom home therapy is standard practice. There is no reason why others with less common factor deficiencies should not give their own treatment, especially if they live in remote locations.

Training for home therapy starts very early in the haemophilic child's life. As he grows, his parents, and later the child himself, learn to recognize which bleeds need treatment and how the relevant clotting factor is prepared and given. As their confidence grows they are able to undertake more and more of the responsibility for direct care until, when they are ready, injections can be given at home.

In some countries this natural progression has given way to the very early prescription of prophylaxis. The idea behind this form of treatment is to prevent all joint bleeds. Prophylaxis is discussed further on p. 166.

Products for home therapy

Freeze-dried (lyophilized) concentrates of factors VIII and IX are most suitable. The vials contain a known amount of the factors and a suitable dose can be prescribed for a particular patient. Many concentrates may be stored at room temperature (up to 25 °C or 77 °F) for up to 6 months, but in general all the products are best kept in a refrigerator at between 2 and 8 °C (36 and 46 °F). In some countries concentrates may still be prepared which, in order to maintain their stability, must be kept in a deep-freeze at around −20 °C (−4 °F). The leaflet enclosed with each vial will contain the information specific to each product. Concentrates are prepared for injection by mixing them with sterile distilled water (diluent) immediately before injection. Diluent does not have to be stored in a fridge.

Wet cryoprecipitate may still be used in a few programmes. Less stable than the dry preparations, it must be stored in a deep-freeze well away from food. The storage life is limited and it is more difficult to draw up before injection. Centres issuing cryoprecipitate or fresh-frozen plasma will teach patients the best methods of administration.

The home-therapy programme

Two aspects of home therapy may worry families offered this form of treatment. The first is the very natural fear of 'the needle'. The thought of performing venepuncture on a son, husband, or friend without immediate assistance is frightening. The knowledge that you are being taught to do it to yourself if you are the person with haemophilia even more so!

Few children like to place a needle in a vein by themselves—that comes in the early teens for most—but all are capable of managing the rest of home therapy with a panache that puts the average doctor to shame.

The second worry concerns loneliness. After a working relationship with haemophilia centre staff the thought of losing contact, despite the independence it implies, brings concern. There should be no need for this because the majority of centres now encourage people to telephone or call in, and regular follow-up has become established. When geography or some other reason intervenes to distance families from a centre, the Haemophilia Society can bring great benefit.

Training for home therapy cannot be performed as a strict and limited regime. Every family is different and some learn faster than others. Training really starts in the course of the very first consultations about haemophilia, continues through the time when parents learn which bleeds to treat and conquer their fear, and culminates in a period of formal instruction. It does not end there because the management of haemophilia is a living thing, a dynamic process in which families and their advisers continually learn to adapt their practices to the best possible benefit of the affected person.

The following instructions are those we use in the Newcastle Haemophilia Centre. Other centres will recommend slightly different regimes, depending on the blood product in use. It is not without good reason that the instructions read like a contract. Without agreement between doctor, patient, and patient's family home therapy cannot work.

Instructions for home therapy

You are about to start treating yourself at home with antihaemophilic factor concentrate. If it is used sensibly, the high cost of this concentrate should be offset by savings in your time, and in ambulance and hospital time. Your training and the following notes will help you to use home therapy safely and effectively. If you experience difficulty, please do not hesitate to contact one of the centre staff.

The kit

The following items are those usually supplied with the concentrate and diluent (sterile water). Modifications will be made to suit individual patients and products.

(1) record-keeping book
(2) dry cotton wool balls
(3) an antiseptic
(4) plastic syringes
(5) filter needles (in concentrate pack)
(6) transfer needles (in concentrate pack)*
(7) small-vein sets ('butterflies')
(8) disposable needles

* These may be single or double-ended.

The kit (cont.)
  (9) needle disposal box
 (10) tourniquet
 (11) paper adhesive tape (Micropore) and Band-aids
 (12) antihistamine (Piriton) for IV injection
 (13) thermometer (if warm water is needed to help concentrate dissolve)

Indications for home therapy

Home therapy should be given in the event of:

 (1) bleeds into joints
 (2) bleeds into muscles
 (3) injury to mouth, tongue, face, eyes, or neck*
 (4) severe knocks to the head or unusually severe headache*
 (5) prolonged bleeding from any site*
 (6) severe swelling in any site*
 (7) open wounds with prolonged bleeding
 (8) following any accident that may result in a bleed.

If you think you are bleeding do not hesitate to administer concentrate; **the earlier bleeds are treated the better**.

Home therapy by itself is suitable for uncomplicated joint and muscle bleeds and for bleeding from cuts which do not require stitches, and for the initial treatment of haematuria. In the event of any other bleed (*see above) you should contact the centre as soon as possible. Remember that for open bleeds in those with haemophilia A an antifibrinolytic is usually indicated as well as blood product, with the exception of haematuria.

Giving the concentrate

The following general instructions cover most concentrates but some differences in preparation may apply. Always check the insert leaflet packed with the product you are about to use, and ask your centre staff for help if you do not understand any of the instructions. Check the expiry date on the concentrate bottles. Do not use if the contents are time expired.

Getting ready

 1. Wash your hands with soap and water.
 2. Wash down the preparation area with detergent and cover it with a protective cloth. Check that all the items you will need are within reach (Fig. 7.1)
 3. Make sure you are in a comfortable position before starting; the intended site of injection should be well supported.

It is wise to have a responsible person within call while you give yourself

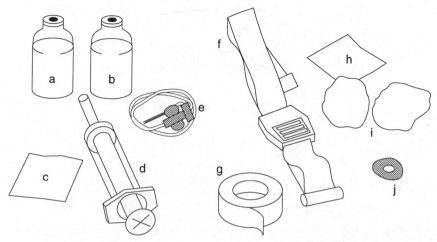

Fig. 7.1. The kit required for the intravenous injection of a factor VIII or IX freeze dried concentrate. (a) sterile water with flip top cap; (b) sterile concentrate with flip top cap; (c) antiseptic swab for wiping tops of bottles; (d) sterile syringe with needle guard in place; (e) butterfly needle; (f) tourniquet; (g) tape; (h) antiseptic swab for wiping skin; (i) cotton wool balls; (j) sticky plaster.

treatment, especially if you are still getting used to the techniques involved.

Disposable gloves should be worn throughout the procedure by anyone giving treatment to someone else. The gloves should be taken off only when all used equipment has been safely packed away. Hands should be washed in soap and water before and after using the gloves. These simple hygienic measures protect both treater and patient from infection.

Preparation

1. Take the protective caps off the concentrate and diluent bottles.
2. Wipe tops of bottles with antiseptic (Fig. 7.2).

EITHER

When double-ended needle supplied in kit:

3. Remove plastic cover from one end of double-ended needle; take care not to touch exposed needle. Insert needle into diluent bottle so that the needle lumen is just inside the cap.
4. Remove other plastic cover from double-ended needle. Invert diluent bottle over concentrate bottle and push exposed needle into concentrate bottle. The vacuum in the concentrate bottle will draw in diluent. Remove the needle from the diluent bottle first so that any air is expelled from the concentrate bottle. Go to 5.

OR

When double-ended needle not supplied in kit:

Fig. 7.2. Wiping the tops of the bottles with an antiseptic swab.

3. Attach the plain needle to the syringe, taking care not to touch exposed needle. Insert needle into diluent and withdraw entire contents into syringe (Fig. 7.3). A few millitres (ml) of air gently pushed into the diluent bottle will aid complete withdrawal of water.
4. Withdraw needle and insert into concentrate bottle; inject all diluent (Fig. 7.4). Withdraw needle, and discard.
5. Gently rock and rotate concentrate bottle until concentrate is dissolved. Vigorous shaking will lower clotting factor activity.

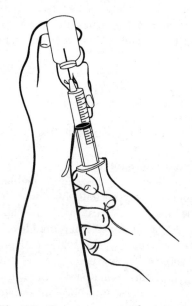

Fig. 7.3. Drawing up sterile water.

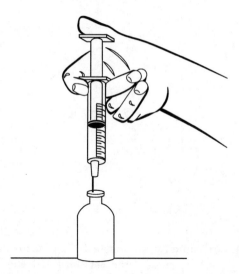

Fig. 7.4. Introducing sterile water into concentrate.

Most modern concentrates go into solution very quickly but if, after 10 minutes, there are still lumps in the bottle, move the bottle in warm water for a few minutes to help dissolve them.

6. Attach filter needle to the syringe; wipe concentrate bottle top with antiseptic; insert needle, invert bottle, and withdraw fluid. It may be necessary to inject some air into the bottle to aid withdrawal. Take care not to touch needle.

7. Remove filter needle and attach small-vein set to the syringe.

8. Holding syringe vertically (plunger down), gently push plunger in until fluid just reaches the tip of the plastic needle guard. This expels the air from the syringe and tubing; the injection of air may be dangerous (Fig 7.5).

Giving the concentrate

1. Sit or lie in a comfortable position. Tear off a strip of adhesive tape ready for holding the needle in place, and put a dry cotton wool ball and a Band-aid near at hand.

2. Apply tourniquet about 10 cm (4 inches) above chosen vein. If you are using a hand or elbow vein hang arm down to allow the vein to fill with blood before tightening the tourniquet.

3. Swab the skin over and around the vein with antiseptic for at least 30 seconds, and then allow it to dry. (Fig. 7.6).

4. Remove needle guard and gently insert needle into vein (Fig. 7.7). The syringe may rest on a table or in your lap while venepuncture is performed. When using an elbow vein keep arm straight at the elbow supported by a pillow. When using a hand vein, support

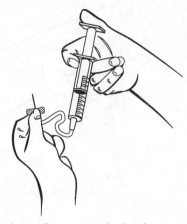

Fig. 7.5. Syringe with butterfly set attached. The reconstituted concentrate should be slowly pushed through the set until it reaches the tip of the needle.

---

hand over knee or table top, and make a fist to tighten the skin and hold vein in place. Push needle well into the vein and secure to skin with adhesive strip. If needle is in vein, blood will flow back and be seen in the tubing. Do not let it flow all the way back up the syringe.
5. Release the tourniquet.
6. Inject fluid (Fig. 7.8). If this is painful, or a swelling appears at

---

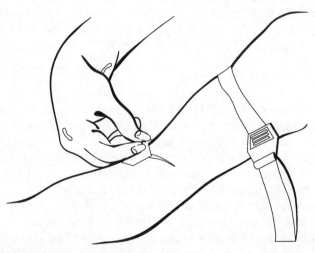

Fig. 7.6. With the tourniquet applied and a vein identified, the skin over the vein is wiped with an antiseptic swab.

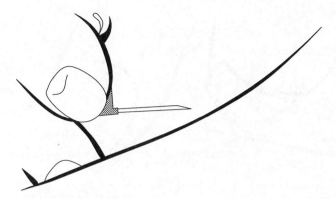

Fig. 7.7. Holding the butterfly wings, the needle is introduced into the vein with the bevel facing upwards.

the injection site, the fluid is entering the tissues and not the bloodstream. Another venepuncture will be necessary if the needle cannot be readjusted in the vein. Do not worry—everyone misses sometimes!

7. If palpitations or other unusual feelings are experienced during the injection, stop and wait for a minute or two before proceeding. The injection rate should be about 3 minutes for each 10 ml of fluid.

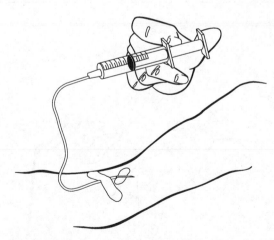

Fig. 7.8. Injecting the reconstituted concentrate. Note that the sticky tape is over the tubing and not over the butterfly itself. This makes for easier withdrawal once the injection has been given. Fixture over the wings of the butterfly is used for more permanent infusions. Note that the tourniquet has been removed before injection.

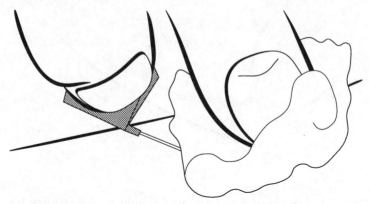

Fig. 7.9. Withdrawing the needle after injection. Once again the butterfly wings are used. As the needle tip leaves the skin, pressure is put on the cotton wool ball.

8. Once infusion is completed place a dry cotton wool ball over injection site (Fig. 7.9). Withdraw needle and apply pressure to site for 2 minutes, keeping arm elevated. This can be done by pressing upwards against the underside of a table (Fig. 7.10). When an elbow vein has been used do not bend the arm because this will kink the vein temporarily and bleeding may restart when the elbow is moved again. When bleeding has stopped, cover the injection site with a sticky plaster (Fig. 7.11).

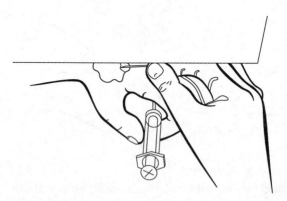

Fig. 7.10. Withdrawal from a hand (or wrist) vein. The puncture site, covered with cotton wool, is pressed on to the underside of a table while discarding the needle.

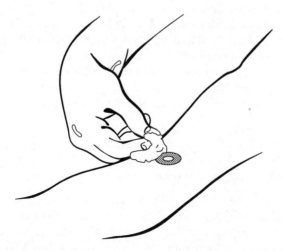

Fig. 7.11. Pressure is maintained over the vein for two or three minutes and then a sticky plaster applied. This can be removed in 24 hours. Note that the arm is straight at the elbow. Trying to seal the vein by pressure with the elbow bent does not always work (see text).

After treatment

When treatment is completed, and you are satisfied that the injection site is sealed and not likely to bruise, the following procedures should be carried out.

1. Place all needles (including small-vein set) in needle disposal box. If the box is large enough it is easier to keep the needle used for giving the concentrate attached to the syringe, and to dispose of both together.
2. Place bottles and dirty dressings (and any remaining syringes) in plastic pack. Seal the pack. If you have a suitable fire, the syringes and dressings may be burnt.
3. Return the needle disposal box and plastic packs to the centre, or to your nearest hospital. Do not put them in the dustbin or refuse sack. If, for any reason, concentrate is drawn up into a syringe and not used it should be returned to the bottle for disposal by the centre.

Danger

There is a very real danger of infection from used needles and other 'sharps'. Both hepatitis and HIV infection have been transmitted to others by needlestab, so the correct handling and careful disposal of needles is extremely important. Needles should NEVER be re-sheathed (put back into their original plastic covers) before disposal because experience has shown that most needle-stick injuries happen when this is attempted. They should be discarded straight into the disposal box, which should never be allowed to overfill. Stop using a box when it is about two-thirds full and seal it for disposal. Never push the contents of a box down to make more room.

However careful people are, needle-stick will happen occasionally in the best regulated homes and hospitals. If the person treating someone else is stabbed with a used needle or other 'sharp' he or she should:

- allow the puncture wound to bleed freely;
- wash the site with plenty of soap and water;
- swab the site with an antiseptic and cover with a Band-aid;
- report the injury to their doctor or to the centre within 24 hours.

Some doctors suggest that zidovudine capsules should be available to take within one hour of deep needlestab or other accident involving the possible injection of HIV infected blood into someone who is not infected. The likelihood of this happening in the course of haemophilia treatment (as opposed to taking blood from someone) is *extremely remote*. Your Centre doctor will have up-to-date knowledge of this suggestion.

Those responsible for treating others with blood products should be immunized against hepatitis B, but hepatitis C and other viruses may still contaminate blood products, even those that have been specially treated.

All equipment must be kept out of reach of children and syringes must **never** be washed out and given to children for use as water-pistols. Even the most careful washing will not remove the viruses responsible for disease. If blood or blood products are split accidentally they should be wiped up with a disposable towel soaked in household bleach (hypochlorite). Syringes, needles, and small-vein sets must never be reused for injection.

Allergic reactions

These are rare with concentrate, but if they do occur an injection of the antihistamine Piriton should be given into a vein, as directed by the centre doctor. The injection is given using a 2 ml disposable syringe and needle. Disposal of the ampoule, syringe, and needle is as above.

Record keeping

Please keep a record of all treatment. Further supplies will only be issued on return of records to the Centre. If for any reason a dose is made up and not used this should be recorded.

Record keeping
The following records should be kept:

1. The product used for treatment
(a) name of product;
(b) amount given (in units of clotting factor);

(c) batch or lot number (on the bottle or pack).

This information is vital if side-effects occur. If it is not recorded it is impossible to link these with specific therapy, and impossible to trace infected donors or faults in manufacture.

2. *The treatment itself*
(a) date and time of treatment;
(b) reason for treatment:
   • prophylaxis
   • on demand: site of bleed
                  : time bleed started
                  : cause if known
                  : if first treatment for this bleed;
(c) name of person giving treatment;
(d) any immediate side-effects (raised temperature, palpitations, sweating, headache, etc.) and how they were dealt with;
(e) difficulties in mixing or giving treatment;
(f) effectiveness of treatment.

This information allows both you and your centre staff to follow the course of haemophilia properly, using accurate, on-the-spot data. From the running record of a particular child or man with haemophilia you and they can see how effective treatment has been over time. Recommendations can then be made which may, for instance, help stop runs of bleeds in a particular joint. The best way of doing this is to plot each injection on a calendar chart (Fig. 7.12). A series of charts provides a running record of haemophilia treatment over the years.

Background information
Dosage

The dose of factor VIII or IX given is always worked out to the number of units in the concentrate bottle. This means that none of the clotting factor has to be wasted. The only exception to this rule is when treating very small babies.

Dosage varies according to several criteria. These include the age and weight of the person with haemophilia, the site and nature of bleeding, how soon treatment is given after the recognition of bleeding, and whether the injection is being given on demand, or as prophylaxis. In addition, different people respond in different ways to the same dose of factor VIII or IX. This means that the best dose for an individual can only really be determined by experience, and sometimes by measuring his response to treatment in the laboratory. For instance, following surgery it is routine to monitor the factor VIII or IX levels for several days, and to tailor doses and the interval between them according to the assay results.

There are formulae for working out dosage. These are usually printed

NAME:

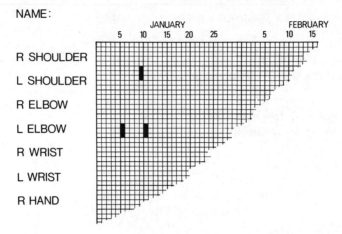

Fig. 7.12. A section of the Newcastle calendar chart for recording the treatment of bleeds. Sites of possible bleeding are listed down the left side. Days of the year run across the top. The site and timing (morning or afternoon of a particular day) of a bleed are taken from the patient's record card or diary. A vertical line is drawn on the calendar, using the top of the appropriate space to show a morning treatment (12 midnight to 12 midday), and the bottom to show an afternoon treatment (12 midday to 12 midnight).

Days in hospital, bleeds caused by injury (trauma), and doses can also be recorded on the chart, as can prophylaxis. Calendar recording gives an immediate visual picture of an individual's progress which is much clearer than trying to decipher a series of written notes or cards. Given records from many patients, the care team can also see when someone is using treatment which is inappropriate for his disorder or his age, and help him to manage his life better.

---

in the data sheets packed with the concentrate. For instance, the formula for one concentrate is given as:

Bodyweight in kg × 0.5 i.u. × desired rise in factor VIII % = Number of factor VIII units required.

The desired rise depends on the bleed. The manufacturer of this product recommends that a rise of 20–30 per cent should be the aim for mild to moderate haemorrhages.

These formulae are intended to be no more than guidelines, but they do give an idea to people with haemophilia who do not have any access to experienced staff.

There are other, rule-of-thumb guides to what dose someone with severe haemophilia should be injecting when he knows he is bleeding. The smallest amount of factor VIII in a bottle is usually around 250 units, and of factor IX 500 units. If *given early enough* this dose stops 80 per cent of acute bleeds. The remaining 20 per cent of bleeds require

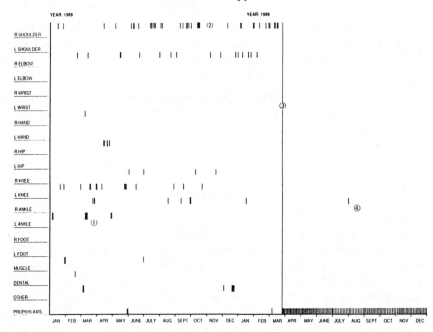

Fig. 7.13. A calendar record of treatment in severe haemophilia A. Each vertical line is a single treatment. Failure of a particular bleed to respond to treatment is easily identified by looking for blocks of lines, for instance in the March 1988 record of a bleed into the left ankle ①. Recurrent bleeding signifies a target joint, which needs to be looked at with a view to a change in management. Thus recurrent bleeds into the right shoulder in particular ②, but also into the right elbow and ankle, and both knees, suggested that regular prophylaxis would help. This was started in March 1989 ③ and only one bleed (right ankle), which responded to a single treatment, occurred in the next nine months. ④ Prior to this regular prophylaxis the patient had given himself injections which anticipated events which he thought might result in bleeding (May 1988, and March 1989). All the rest of this therapy was 'on demand', in response to his recognition of bleeding.

---

further injections, most stopping with one more dose and all after four doses. The key is usually the timing of the treatment rather than the size of the dose.

Prophylaxis often works with similar small doses given regularly. I usually start someone needing prophylaxis at a dose of 500 units of factor VIII on alternate days. If this stops all bleeding either the dose can sometimes be halved or the time between doses extended. If bleeding continues the dose is raised accordingly. People with haemophilia B on prophylaxis generally require less frequent injections than those with haemophilia A because of the longer half-life in the bloodstream of factor IX.

If the aim of treatment is the *total prevention* of arthritis in adult life research has shown that the doses of clotting factor prescribed have to be much higher than the 250 units minimum. In addition, it is now recommended that they be given as prophylaxis (see text).

A very useful guide to the best doses for specific bleeds is *Haemophilic bleeding: early management at home*, by Dr Tony Aronstam, published in 1985 (ISBN 0–7020–1084–7) by Baillière Tindall.

Storage

The factor concentrate should be kept in a cool place out of direct sunlight or in a refrigerator (not in a deep-freeze). The insert leaflet will give specific recommendations for the particular product you are using. If you are travelling, it may be taken in a cool container. Provided the seals on the bottles are not broken and the bottles are intact, the packs of concentrate may be stored in an ordinary domestic refrigerator without danger to others. The diluent water bottles and other equipment may, of course, be kept at room temperature. Used bottles or other equipment must never be allowed to come into contact with food because of the danger of viral transmission.

Prophylaxis

There is now firm evidence that prophylaxis given throughout childhood and adolescence prevents haemophilic arthritis. This is not surprising because we already know that youngsters with mild or even moderate haemophilia do not have spontaneous joint bleeds and do not have arthritis in adult life. Prophylaxis simply provides for a constant level of factor VIII or IX in the bloodstream of someone with severe haemophilia.

In order to be sure of success with prophylaxis it is logical to start the injections before the first haemarthrosis, and not to stop them until growth has ceased and the joints are stable. This means starting at around the age of 2 years and continuing to 18 years. No one yet knows if it is necessary to prolong prophylaxis beyond this. Before considering the advantages and disadvantages of this approach to haemophilia treatment let's look at some definitions.

We already know that on-demand (crisis) therapy is treatment given on the first evidence of a bleed, usually as home therapy. Prophylaxis *is the intravenous injection of clotting factor in anticipation of, and in order to prevent, bleeding.* It can be given as:

- *Single dose prophylaxis.* This is an injection given prior to an event that may either result in bleeding, or be disrupted by it.

Someone taking an examination or about to participate in sport may give himself an injection beforehand.

- *Limited period prophylaxis.* This means regular injections given over a limited period in order to reduce the incidence or frequency of bleeding. Reasons for this type of prophylaxis range from the need to help target joints stop bleeding and settle down, to cover for the honeymoon or holiday.
- *Long-term or permanent prophylaxis.* It is this type of therapy that is intended to prevent arthritis. It is sometimes called maintenance therapy.

In addition, some doctors use the term prophylaxis when prescribing cover for physiotherapy after surgery, or as a precaution in children who have experienced bleeding into the head.

Prophylaxis requires a guaranteed supply of factor VIII or IX concentrate, good venous access, and someone prepared to give the injections. The timing of the injections is usually tailored to the individual but, in general, those with haemophilia A need three injections a week and those with haemophilia B one injection a week. The difference is due to the longer half-life of factor IX. The factor VIII injections are given either on alternate days, or on Mondays, Wednesdays, and Fridays, leaving the weekend clear. Some people like a slightly larger dose on the Fridays, especially if 'weekend' means increased physical activity.

Prophylaxis cannot be given in a vacuum, divorced from all the realities of haemophilia. Attention to detail and regular follow-up are essential parts of any prophylactic programme. Suggested criteria for prophylaxis in childhood are:

- The haemophilia must be severe (factor VIII or IX < 1 per cent).
- The programme must be acceptable to the family.
- The person with haemophilia and his parents should know when to treat a bleed.
- The family should have a commitment to the comprehensive care of haemophilia, which includes accurate recording of treatment and follow-up.

Evidence of severity can only come from a family history, early bruising, and the measurement of factor VIII or IX in the laboratory. No one can know whether or not the individual embarking on long-term prophylaxis would have been prone to recurrent joint bleeds. Long-term prophylaxis has to be started on the probability that this would be the case. This is one of the problems with starting early. Present evidence suggests that 10 per cent of children with severe haemophilia do not develop arthritis anyway. So, if long-term prophylaxis becomes the usual way of treating haemophilia in childhood, 10 out of every 100 youngsters may be subjected to 16 years of injections without benefit.

Another problem is that by starting the programme so early children grow up without being able to recognize the early warning signs of bleeding. This lack of recognition could be disastrous in later life. If a major joint bleed occurred because of failure to treat early, much of the preceding prophylaxis may have been in vain. This is one of the arguments for perhaps continuing a prophylactic programme throughout life, rather than stopping it at 18 years of age.

An emotional acceptance and relaxed relationship between family and haemophilia centre staff will help people accept the potential value of prophylaxis. Without this the discipline needed for prophylaxis cannot be present. It really needs a tremendous commitment to embark on the regularity of treatment demanded by long-term prophylaxis. The situation is not quite the same as that experienced by families with children with diabetes. They know that failure to inject insulin properly would be life threatening. The family with haemophilia—and, more especially, the bored teenager—know that most bleeds are not life threatening and respond readily to treatment.

The cost of prophylaxis

On average the person with severe haemophilia bleeds 35 times a year. Bleeding is likely to be more frequent in childhood, and it is not unusual for a higher dose of factor VIII or IX in terms of body-weight to be required in order to stop bleeds in children. Overall, the average boy with severe haemophilia requires around 50 000 units of factor VIII a year when he is using 'on-demand' therapy. On prophylaxis at the lowest dose of 250 units (one vial of concentrate) three times a week he will need 39 000 units *plus* treatment needed for any break-through bleeds. In practice, the optimum dose for prophylaxis is probably 2000 units per kilogram of body-weight a year. This means that:

- at 5 years the annual requirement for factor VIII is 40 000 units;
- at 10 years it is 60 000 units
- at 15 years it is 110 000 units; and
- at 18 years it is 140 000 units.

Overall then, more clotting factor concentrate is required for long-term prophylaxis than for a home-therapy programme reliant on 'on-demand' treatment. However, in terms of a lifetime the cost of early prophylaxis may be offset by less need for orthopaedic surgery in adulthood, and less of an infringement on the quality of life in terms of pain, reliance on others, and absences from school or work.

Recommendations on prophylaxis in the United States of America

In 1994 the Medical and Scientific Advisory Council of the National Hemophilia Foundation issued the following recommendations about prophylaxis. Similar recommendations are already in use in some other countries. I have adapted them here for guidance, with the permission of the Foundation.

'In view of the demonstrated benefits of prophylaxis begun at a young age in children with hemophilia A and B, we recommend that prophylaxis be considered optimal therapy for children with severe hemophilia A and B (Factor VIII or Factor IX < 1 per cent).

Prophylactic therapy should be started early (beginning at 1–2 years of age), with the aim of keeping the trough FVIII or (FIX) level above 1 per cent between doses. (This can usually be accomplished by giving 25–40 FVIII units/kg three times per week, or 25–40 FIX units/kg twice weekly.)

We also recommend that regular follow-up be performed in order to evaluate joint status, to document any complications, to document all costs associated with each child's prophylaxis and to record any bleeding episodes which occur during prophylaxis.

Prophylaxis should be *considered* for other age groups as well. In as much as the economic benefits realized could be substantial, a cost–benefit analysis should be included as part of the assessment.

The National Hemophilia Foundation encourages persons with hemophilia and their families to consider these recommendations within the context of comprehensive care, which emphasizes patient and family involvement in the decision-making process, based on a thorough discussion of the risks and benefits—medical, social, psychological, and economic—of such care. Thus, in considering prophylaxis, the NHF encourages persons with hemophilia and their families to examine the following issues with their medical team:

A. For parents of young children (aged 1–2 years) with severe hemophilia who are being considered for prophylaxis.

- Risks versus benefits of prophylaxis
  - need for frequent clotting factor infusion versus less joint damage;
  - potential psychological implications;
  - potential costs and reimbursement implications;

    – possible venous access problems, necessitating use of a central venous access device (such as a surgically implanted port);

    – potential complications of such central venous access devices (examples are bacterial infection, often necessitating hospitalization for intravenous antibiotics; clots in central line; may need to be removed and another line reinserted later);

    – possibility of other complications not yet identified;

- Type of products most appropriate for prophylaxis, including current knowledge of infectious risk of products. Some believe that ultrapure plasma-derived and recombinant products would be most appropriate for prophylaxis because of possible decreased infectious risk.

B. If prophylaxis is being considered for an older child, adolescent, or adult

- history of patient's clotting factor usage and how prophylaxis would compare to that usage;
- current joint damage and what can reasonably be expected from prophylaxis;
- those considerations listed under A, above.

C. If prophylaxis is selected as the treatment of choice, then the responsibilities and understanding of potential risk for the individual or families need to be clearly understood. The following points should be included in discussions held with people with hemophilia and their families.

- frequency of infusion;
- potential psychological implications;
- standard care for the central line that the family must follow;
- frequency of comprehensive follow-up;
- statement of indications to discontinue;
- occurrences that prompt an immediate call to the physician or nurse;
- essential need for regular follow-up;
- a statement of continued risk of bleeding with risk-taking behaviour;
- an acknowledgement that there may be current risks not yet identified.

After a thorough discussion with their medical team, people with hemophilia and their families should decide which procedure they should follow. This decision should be evaluated periodically, particularly in light of changes in bleeding and clotting factor usage.'

Venepuncture—more about technique
Choosing the vein

Easily accessible veins are those on the back of the hand, the front of the wrist, the outside of the forearm above the base of the thumb, and in the elbow crease. Those on the back of the hand and at the elbow are the best. They are easily seen and felt, and are long and straight, and it is easy to apply pressure to prevent bruising after treatment. The veins are brought into prominence with the use of the tourniquet. If the arm is allowed to hang down for a minute or so before application of the tourniquet blood will start to pool and the veins will fill better. Immersion of the hand in hot water, and rubbing and flicking the skin over a vein will also help to bring it into prominence. Always try to treat yourself in a comfortable position; anxiety causes veins to contract. A hot bath may help before prophylaxis. Relax and let your veins relax with you!

Provided precautions are taken to avoid infection and bruising there is no reason why the same vein should not be used for every injection. It becomes a 'mainliner' and technique becomes automatic and pain-free with time. However, it is wise to know of more than one site for venepuncture so that if your favourite vein is in the limb affected by a bleed another can be used. I *do not* advise the use of a local anaesthetic cream (lignocaine and prilocaine, Emla cream) before venepuncture. Staff at my centre have found it both unnecessary, and likely to make veins more difficult to find after it has been rubbed into the skin over them.

Entering the vein (venepuncture)

This is surprisingly easy and pain-free with modern needles, and children with haemophilia soon get used to it and are not frightened.

Keeping the bevel up (Fig. 7.7), the needle is slid through the skin with a firm controlled movement. It should enter the skin at an angle of about 30° and at a point which will allow the fully placed needle to lie on a firm surface. Once through the skin the axis of the needle should be lined up with the vein, and at only a slight angle to the skin. A second purposeful movement will slide it into and up the vein. It is preferable that at least 1 centimetre of the needle shaft should lie in the vein to prevent it slipping out during injection.

Withdrawal from the vein

If pressure is not applied over the venepuncture site immediately *after* withdrawal a bruise will rapidly appear, especially if the arm is in a bent position. Heavy pressure over the needle before withdrawal may damage the vein and should be avoided.

If an elbow vein has been used a cotton-wool ball is put over the site following withdrawal, and the elbow flexed while the needle is discarded. The elbow *should then be straightened* and pressure applied with a finger, directly over the wool for at least 2 minutes. If the elbow is kept flexed, the vein may simply be kinked and will open up again on movement.

Care of the veins

The veins are the lifeline of someone with haemophilia. Modern equipment allows venepuncture with only the very slightest of undue injury to a vein, and this should be temporary provided proper hygiene has been observed. Veins should only be exposed surgically (the technique of 'cut-down') in cases of extreme emergency. This is because during the cut-down procedure part of the vein has to be tied off, meaning that it cannot be used again. Some of the older generation of those with haemophilia have few veins left for venepuncture because of cut-downs in their childhood. When this interferes with treatment a shunt may be indicated. A shunt is created by surgically joining an artery to a vein. The pressure of blood entering the vein expands a length of it, making it an obvious and easy site for venepuncture. An alternative is a PortaCath, described at the end of this chapter.

The major sites of veins suitable for venepuncture have already been discussed (p. 168). In fact any vein will do. In the small infant scalp veins are often used. In adults the veins over the foot are easily accessible. Whichever vein is used, always remember to release the tourniquet before injection (if you don't the vein will burst), and to keep pressure over the site of venepuncture for some minutes after injection. If a vein becomes sore or looks inflamed it should not be used until it has recovered.

A useful method of practising venepuncture is to put a length of plastic Butterfly tubing under a tablecloth and attempt to needle it. This saves relatives, friends, and staff from multiple stab wounds!

Veins are more difficult to find in the obese and in the patient with poor muscular development. Dieting and repeated exercise with a spring-loaded grip exerciser or hand sponge will benefit these people.

Healthy veins feel soft and bouncy, and refill quickly after they are depressed. Dry skin, discoloration, and scarring over sites used for repeated venepuncture may be made less obvious by rubbing lanolin cream into the area regularly.

Hit and miss

Everybody, even experienced haemophilia centre nurses (and occasionally doctors), misses a vein sometimes. When this starts to happen more often than you would like, you need to take a rest. Respite care for you (and the patient) by transferring injections to the centre or family doctor, or by coming off prophylaxis, for a month or so may help. The hits will return!

Venous access when things get really difficult

Veins in small people sometimes prove troublesome, both to find and to get into! Thankfully babies with haemophilia do not need treatment very often and, when they do, there is usually a vein which can be used at least once. In infants without haemophilia the veins of the neck or groin are sometimes used but when the child has a bleeding disorder this is a dangerous practice. Venepuncture at these sites can result in very nasty bruising.

When treatment into a vein is needed regularly it is sometimes advisable to insert an artificial line or catheter. This is done under anaesthetic. One end of the catheter lies in the right atrium of the heart. The other end, which provides access for the injection of treatment, is either implanted under the skin of the chest wall or is taped to it. There are many names for these devices. That with the end implanted under the skin is usually called a PortaCath; there are similar devices with different trade names. The others are referred to as long lines; two of the trade names are Hickman or Broviac catheters.

These catheters were developed originally for people with kidney failure who needed regular blood dialysis, and for people with some sorts of cancer who needed regular chemotherapy. They have proved to be very successful, but they do need to be looked after very carefully. Two of the problems associated with them are infection and blockage due to clotting.

For young children with haemophilia and poor veins who are on prophylaxis a PortaCath is probably the best device. It ends below the chest skin in a metal button with a latex rubber membrane through which a needle can be inserted many times. It is put into place through a small cut near the right nipple and another cut through the skin of the neck. The soft, flexible catheter is introduced into one of the neck veins and gently pushed down this vein into the right atrium. The other end of the catheter is then tunnelled under the chest skin to join up with the button. The skin cuts are then stitched and once healing is complete there is hardly anything to see except the bulge over the button. Activity is not restricted. Because the device is completely buried the risk of infection is less than with catheters which end outside the skin. If looked after properly a PortaCath should last for many months or even years. With luck only one insertion will be needed before the child's other veins have developed enough to allow prophylaxis to continue. When it is no longer needed the catheter and its button are removed. There are no long-term after-effects.

More about the use of PortaCaths in relation to prophylaxis will be found on p. 167.

A Reminder
The rules of haemophilia treatment

1. THE EARLIER THE BETTER
   Early treatment of a bleed prevents later damage. The more blood that is allowed to enter a joint or muscle, the greater the subsequent damage to the tissues, and the longer the time taken for recovery. Early treatment usually allows an immediate return to shool or work, and also diminishes the chance of arthritis and disability later in life.

2. IF IN DOUBT, TREAT
   Trust your (or your child's) 'aura'. If you feel that a bleed might have started, treat it. NEVER wait until a joint is hot, swollen, and painful. Do not worry that you may 'waste' the occasional treatment by injecting when a bleed is not present.

3. A SHOT IN TIME SAVE VII (or IX)
   In general early treatment saves blood product. A small dose of factor VIII or IX stops small, early bleeds. Bleeds left to develop require more, and often repeated, doses of blood product to stop them.

8

Physiotherapy

The maintenance of healthy joints and muscles is crucial to the quality of life of the person with haemophilia. Clotting factor replacement may stop or even prevent bleeds, but it does not restore joint or muscle function. Only regular movement can do that.

If a house is neglected things are far more likely to go wrong and be harder to put right than if the property had been kept in a good state of repair in the first place. The same goes for the body. A well-maintained body shrugs off everyday stresses and strains, and repairs itself far more quickly than a body which has been neglected.

The task of the physiotherapist or physical therapist working in a haemophilia centre is therefore three-fold:

(1) to encourage maintenance of health;
(2) to teach people how to prevent problems occurring in the first place; and
(3) to speed repair when things go wrong.

Most of this chapter is concerned with the therapy—putting things right, because the usual narrow, hospital definition of physiotherapist is 'the art of restoring natural function after injury or disease'. The broader health maintenance approach is considered further in Chapter 14.

The techniques employed by the physiotherapist depend on a profound knowledge of the mechanics of the body—how joints, bones, and muscles work on principles of leverage and nervous control—as well as a compassionate understanding of the needs of the injured person. The skills of the physiotherapist can be of enormous benefit in two ways. He or she knows how to bring joint and muscle function as near to normality as possible and as quickly as possible after acute bleeding episodes, and can give much useful advice on how to maintain mobility after discharge from hospital. In contrast to many forms of medical treatment, in which the patient may feel he is an inanimate object, physiotherapy can work only with the full participation of both teacher and pupil.

Because the needs of each patient differ it is not possible to give a step-by-step guide to physiotherapy techniques in a book. What follows is a general guide to the management of some of the more common bleeding episodes and some suggestions for exercises to perform at home.

Acute bleeding episodes

With haemarthroses, the immediate need is for factor replacement *as soon as possible* after the start of the bleed. Residual pain may be relieved by the application of ice, but care must be taken when this is done in order to avoid ice burns. Crushed ice must be wrapped in a wet towel (ice cubes can be broken up inside the towel with a hammer) before application. If a plastic pack containing crushed ice is used, the skin should be treated with oil first. Skin protection is very important when either cold or heat application is carried out after surgery. The skin area involved may have been affected by temporary loss of sensation and nasty burns may result.

For patients with high-titre inhibitors (Chapter 10), and for people with a haemarthrosis or deep muscle bleed which is particularly severe or slow to resolve even with factor replacement, physiotherapy is necessary as soon as bleeding has stopped. In these situations the joint and its associated muscle groups will have been immobile for a day or two, and will be stiff and weak.

The aims of treatment are to restore muscle power, and to restore the range of movement of the joint. If muscle power is not restored, the joint will remain unstable and at risk from recurrent bleeds. One form of treatment which is especially useful in helping to speed up the resolution of deep muscle bleeds is pulsed short-wave diathermy. This is used with deep frictional massage followed by static contractions and gradual stretching exercises. The techniques are relatively painless, can be applied immediately, do not require hospital admission, and have no adverse effects on the patient's haemophilia.

Relief of pain without medication

Physiotherapists are skilled in helping people with arthritis overcome chronic pain, and their methods can be of great benefit to people with haemophilic arthritis. The use of ice packs for acute pain has already been mentioned. A wide variety of reusable packs is available from sports shops. More often though, heat is used to help people with chronic pain. Heat relieves muscle spasm, which adds to the discomfort of an arthritic joint, as well as working on the joint itself.

Heat is applied with ordinary hot water bottles or hot packs, suitably wrapped to prevent burns, or by short-wave diathermy which generates deep heat in the tissues. People with painful hands or wrists are sometimes treated with hot wax baths. Massage can be especially valuable when muscle spasm is present.

Another method of achieving both acute and chronic pain relief is with a TENS (transcutaneous electrical nerve stimulation) machine. When injury occurs, nerves from the site transmit pain via relay cells in the spinal cord to the brain. At the same time, other nerves which have been stimulated by the injury also transmit to the relay cells and start to damp down the pain. Thus our perception of pain occurs as the result of a balancing act between these two stimuli. This balance has been likened to a gate: open, it allows pain through; when closed pain is blocked. TENS stimulates the second sort of nerves and, by altering the balance, closes the gate. Modern TENS machines are small boxes which can clip to the belt, with skin electrodes that are placed on or near the site of the pain. The machine is powered by small batteries, and can be used at home.

Pain relief may also be achieved by a similar mechanism with acupuncture. People with a bleeding disorder who want to try acupuncture are advised to talk to their haemophilia centre doctor first. Although most, including those with severe haemophilia, will not be harmed by this procedure when it is performed by a reputable practitioner, it could provoke bleeds in a few people.

Exercises

Readers are referred to an excellent booklet, published by the UK Haemophilia Society in 1993, *Joint care and exercises*, ISBN–1–872945–023, which details exercise for children and adults with haemophilia.

The knee

The major muscle group involved here is the quadriceps or 'quads' (p. 103). The power of the quads muscles is lost extremely quickly if they are not exercised. The following exercises will help strengthen the muscles after injury. After a particularly severe bleed or an operation on the knee joint, the advice of a member of the medical team should be sought before attempting any of the routines described. As with any physical training programme, it is sensible to start with the least demanding routine and work one's way up.

Exercises to strengthen muscles are:

- either static (performed without moving the joint), or
- active (in which one or more joints are moved).

Passive exercises, in which another person moves the joint, should never be practised without medical advice as they may result in further injury.

1. *Static quads*

This is an easy routine to carry out as soon as the acute pain of a haemarthrosis has resolved. During the exercise, the quads muscles tighten and, because the kneecap lies in the tendon of this group of muscles, it can be seen to move up towards the trunk as the exercise is performed.

- Sit or lie down on your back on a bed or couch, or on the floor with the whole of the affected limb supported.
- Press your knee down into the bed.
- Pull your foot up, toes towards your head.
- Hold to a count of 5 seconds.
- Relax.
- Repeat in runs of 5 to start with, and build up gradually to runs of 20.
- When you can do this comfortably 4 times a day move on to the next exercise.

2. *Active quads*

(a) In this exercise (Fig. 8.1), you may need support from a friend if you are very weak.

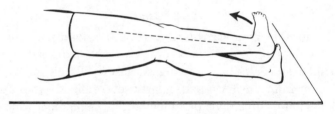

Fig. 8.1. Active quads—side lying—affected leg uppermost.

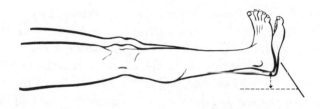

Fig. 8.2. Active quads—back lying—straight leg raising.

- Lie on your side with the affected limb uppermost.
- Holding your knee tight and straight, pull your foot up, toes towards your head.
- Slowly move the limb forwards from your hip for about 30 cm (12 inches).
- Return to the start position, legs together.
- Relax.
- Repeat in runs of 5, building to runs of 20, 4 times a day.

If you are too weak to hold your leg away from the bed get a friend to place a hand, palm up, under your ankle and gently support it during the exercise. When you are strong enough move on to (b).

(b) Straight leg raising (Fig. 8.2)

During the first exercise the quads did not have to work against gravity. This exercise starts with the resistance of gravity and may be built up to a regular routine as part of your fitness programme. During the exercise it is important to keep your big toe pointing towards the ceiling. If you rotate your leg to the side, muscles other than the quads take over some of the forwards movement. This may make things easier, but it is cheating!

- Lie on your back.
- Press the back of the knee down into bed.
- Pull foot up, toes to ceiling.
- *Keeping your knee straight* raise the affected limb so that the heel is 5–10 cm (2–4 inches) from the bed.
- Hold to a count of 3 seconds and lower *slowly*.
- Relax.
- Repeat in runs of 5, building to runs of 20 with increased lengths of hold time, four times a day.

No advantage is gained by increasing the height of lift during this exercise. If the knee is bent the quads do not work properly.

In order to build up the routine, increase the resistance by weighting your leg (Fig. 8.3). The simplest way to do this is to lift a pillow placed over the ankle, and then to graduate to a small sandbag weighing 250 grams (about 1/2 lb) or more. The physiotherapist will usually be able to lend a sandbag for use at home, or you can make one from canvas or close-weave cotton. The bag is held on the sole of the foot with Velcro loops or tape. A thick sock worn under the strapping increases comfort.

Ankle and wrist weight bracelets can be bought quite cheaply from sports shops.

3. Active quads with knee bending

(a) This is a useful first exercise to perform when a knee has been immobilized for a time. It is like exercise 2(a) in that it encourages muscle movement without gravity.

- Lie on your side with the affected limb uppermost.
- Slowly bend the limb at the hip *and* knee, and draw the knee up towards your chin.
- Slowly straighten and return to start position. As you straighten keep your foot pulled up hard at the ankle.
- Relax.
- Repeat in runs of 5, building to runs of 20, 4 times a day.

(b) An important part of the quads group is the inner thigh muscle (called vastus medialis). This muscle helps lock the knee in extension during weight-bearing. In order to exercise it:

- Lie on your back with a rolled-up towel beneath your bent knee.
- Raise your lower leg from the knee.
- Hold to a count of 5 seconds and lower *slowly*.
- Relax.
- Repeat in runs of 5, building to runs of 20, 4 times a day.

(c) When knee flexion (bending) permits, the following exercise (Fig. 8.3) should be tried.**It is not recommended for people with arthritis** as it may aggravate their symptoms, which include pain.

- Sit on a couch or surface which is high enough to allow your feet to clear the ground when your knees are bent to 90°. Your thighs should be supported throughout their length.
- Straighten the affected leg at the knee.
- Hold for a count of 3 seconds and lower *slowly* to the start position.
- Relax.
- Repeat in runs of 5, building to runs of 15–20, 4 times a day.

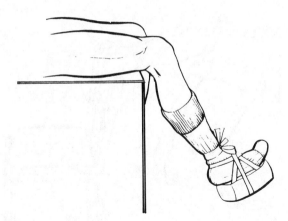

Fig. 8.3. Active quads with bent knee—weight attached.

Progression of this exercise, which should eventually form part of your body maintenance programme, is achieved both by gradually increasing the number of lifts, and by adding resistance to the ankle using weights. Start with 450 grams (1 lb) and build up gradually.

(d) Partial weight-bearing

This is best achieved in a swimming or hydrotherapy pool, the buoyancy provided by the water taking some of the body-weight off the injured limb.

Hydrotherapy is very useful during earlier stages of mobilization and for people with painful arthritis. At first buoyancy is used to assist movements. Later the water is used to provide resistance.

Air rings may be used to support a weakened limb and also to assist specific movements, the buoyancy they provide helping weakened muscle groups. As strength improves air rings can be used to provide resistance to specific muscle groups. For instance, with an air ring placed round the ankle the action of knee bending when floating on one's back becomes very much harder to perform.

Resistance may be increased further by wearing a flipper or, if an arm is affected, holding a table-tennis bat. If a pool is not available, initial attempts at weight-bearing require support from furniture, bars, crutches, or other people. These initial attempts should always be made with supervision.

(e) Full weight-bearing

Once you are able to stand without support this exercise (Fig. 8.4) is a useful one to both increase quads strength and help stretch the calf muscles and Achilles tendons. Try to keep your heels on the ground during the exercise.

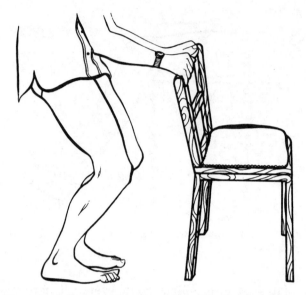

Fig. 8.4. Full weight-bearing—knee bend and straighten.

- Stand facing the back of a chair or a cool wall-mounted radiator.
- Holding on to this support, and with your feet together about 70 cm (2 feet) away, slowly bend your knees as far as possible.
- Hold for 5 seconds and slowly straighten again.
- Relax.
- Repeat as above.

Multigym exercises under supervision or using a programme recommended by your physiotherapist may be introduced at this stage.

4. An exercise for the hamstrings

The bulky group of muscles behind the thigh, the hamstrings, extend the hip and flex the knee. They are partially exercised in most of the routines described above. The following exercise keeps them in trim, and helps prevent tightening of the tendons which can prevent full extension of the knee.

- Lie on your front, legs straight and supported on a bed, couch, or floor.
- Slowly bend the affected knee to 90°.
- Hold to a count of 5 seconds and lower *slowly* to the start position.
- Relax.

• Repeat as before.

As with exercise 3(c), this routine should be built up using increasing weights applied to the ankle.

There are many more active ways of exercising the quads and the other muscle groups of the lower limbs. They include swimming and cycling—two of the best all round sports for body maintenance—walking, skipping, aerobics, and working with weights and springs. The multigym combines many of the latter exercises and is recommended—but see p. 290. Other types of apparatus are also discussed on this page.

The ankle

Following ankle injury or bleeds into the calf muscles, it is very important to ensure that the foot can eventually be pulled up beyond a right angle (90°) with the leg (Fig. 8.5). If this is not achieved, the heel will not touch the ground when standing, and walking will be performed on the base of the toes. The result is a permanent limp.

The muscles of the lower leg and foot are easily exercised by performing all the possible movements of:

• flexion (pulling the foot up);
• extension (pushing it down);
• inversion (turning the foot in);
• eversion (turning the foot out); and
• bending and stretching the toes.

Initially, the movements should be carried out with the foot supported on the bed. With increasing strength, they are performed against resistance provided by the floor, wall, or hands. Exercise 3(e) is a good one for the ankle joints, providing the heels are kept on the floor. Another

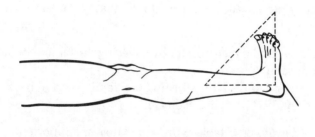

Fig. 8.5. Foot pulled up beyond angle of 90°.

is to stand on one leg with your eyes closed—the art of balancing brings other muscles, including those of the foot, into play.

Early exercises will be made more comfortable if muscles are relaxed by the application of heat. The easiest way of doing this is to have a hot bath, or to soak the feet in a basin of warm water, first.

The upper limb

The most frequent residual disability here is failure to extend the elbow to its normal limit. With frequently repeated haemarthroses, supination (the act of turning the palm up) and pronation (the act of turning the palm down) are also affected. Following bleeds into any part of the upper limb, all the joints should be exercised.

The shoulder joint is of the ball and socket variety, allowing the arm to move in a series of vertical and horizontal arcs, and to rotate in its socket. In health, the forearm can be placed behind the small of the back or behind the neck, and the arm may be swung freely to the side and front to the vertical position. To the rear, an angle of about 70° from the vertical is usual. The elbow should extend and flex fully, and supination and pronation should be unhampered. The wrist extends, flexes and moves from side to side, and the hand and fingers are capable of numerous fine movements; the thumb is vital to hand function. Practise all these movements regularly.

The elbow

Following a bleed into the elbow joint, it is important to restore movement as soon as possible. Where there is loss of movement, friction-free exercises without gravity should be carried out first.

- Sit sideways to a table with your forearm supported on the table-top. Your palm should be flat and facing down.
- Slide your palm across the table and back again. Make sure you keep your elbow on the table so that your shoulder does not move.

This exercise can be made easier by covering the table with talc or flour in order to reduce friction.

Pronation and supination (turning the hand over and back again) can be helped by the following exercise:

- Sit sideways to a table with your forearm supported.
- Rotate your palm upwards and try to touch the table-top with it.
- Add resistance to pronation and supination by holding a rod (a

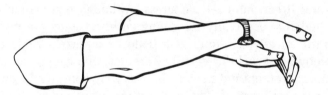

Fig. 8.6. Pushing the fingers and wrist back gently after a forearm bleed.

wooden kitchen spoon will do) in your hand with one end pressed against the table-top as you try to rotate.

The elbow joint is very complex and easily damaged. It should *never* be forcibly straightened, but like other joints can be kept more mobile if the muscles and tendons associated with it are kept healthy. In the upper arm, the two major muscles are the biceps (so-called because it has two heads or attachments) and the triceps (with three heads).

Exercise the biceps by:

- Bending the elbow against resistance provided by the other hand pressed against the front of the wrist, or by
- Bending against the resistance of a weight held in the hand, or a spring.

Exercise the triceps by:

- Straightening the elbow against resistance provided by the other hand pressed against the back of the wrist, or by
- Doing press-ups or working against weights or springs.

Forearm muscles

The most serious bleed of the upper limb is that into the forearm muscles. Untreated, this may lead to a form of claw hand, a permanent deformity. In the days following a bleed, when pain has abated, push the wrist and fingers back gently with the other hand several times a day until full function has been restored (Fig. 8.6). Once again, immersion of the hand in warm water will help.

Home exercises

The importance of home exercises, like those suggested in this chapter, cannot be overstressed. Everyone who wants to remain healthy should take some form of regular exercise, and exercise is vital to people with haemophilia. It helps protect joints by increasing muscle bulk and power and it can help relieve stress. Although regimented exercises can be tedious, especially for young children, they are necessary at least until an affected limb gets better. Exercise periods are made more enjoyable for children if they are part of a game, and especially if parents or brothers and sisters join in as well. Families are encouraged to join in the activities of their local sports centres and swimming pools. Whenever in doubt about exercise ask the advice of the physiotherapist at the haemophilia centre. Physiotherapists can help people with haemophilia. In turn, they can help physiotherapists by exercising regularly!

An easy and inexpensive aid to home exercises is the elastic Theraband. This is graded in a number of colours each providing a different level of resistance. The Armour Pharmaceutical Company provide a similar device called a Flexercise kit which is free of charge to people with haemophilia.

Appliances

These range from simple devices to protect the joints from injury to means of transport. Thankfully, the designers of appliances have at last started to utilize lightweight modern materials, and are giving more thought to the patient as a person rather than as a purely mechanical device. All large sports shops now stock a range of joint supports, designed to protect people with previous problems in a wide variety of sports.

Devices to protect joints

The joints most frequently subject to injury and therefore to haemor-rhage are the knees, elbows, and ankles.

Knees

Whilst there are protective devices that help prevent twists and sprains especially during sports, most do not leave knee movements unre-stricted. In children, however, a few bleeds follow knocks and kicks on the sides and front of the joint. A two-way stretch tubular stocking (Fig. 8.7), padded with foam rubber, will help prevent this sort of injury. The padding is cut out and stuck to the stocking as shown in the diagram. It does not restrict movement and the stocking is unobtrusive and does not show under long trousers. It need be worn only during the day.

With increasing interest in sports injuries and their prevention have

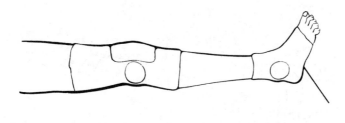

Fig. 8.7. Two-way stretch stocking with inserted foam rubber cut-outs gives protection to the knee or ankle.

come welcome developments in the functional design of splints and other appliances. These are usually tailored for specific injuries or movements; for instance 'active' knee splints are often seen on the professional tennis circuit, or on the ski slopes.

Some people find that knee bleeds commonly start during the night, perhaps following awkward movements during sleep. Runs of these bleeds may occur at frequent intervals. A simple answer is to wear a lightweight plastic splint that holds the knee straight during the night. The splint is taken off in the morning and the joint used normally during the day.

Ankles

People experiencing frequent ankle bleeds, especially children, will benefit from wearing boots or good quality trainers with sides high enough to cover the joints, instead of shoes. The boots support the ankles and protect them from sprains and blows. A similar effect may be achieved using a plastic ankle splint, but this is less acceptable to the fashion-conscious youngster. However, when ankle bleeds are very frequent and not responsive to clotting factor replacement Aircast splints may be very effective.

We recommend that *everyone* with haemophilia should wear shock-absorbing (Sorbothane) insoles and heels in their footwear. Developed for joggers, these simple appliances reduce the impacts of everyday movements on the ankles, knees, hips, and spine.

Elbows

The elbows are injured most frequently by knocks during school activities, and bleeds sometimes follow prolonged desk work when the joints have been in contact with a table top. A padded two-way

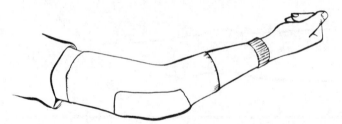

Fig. 8.8. Two-way stretch stocking with inserted foam rubber cut-out protects the elbow.

stretch stocking (Fig. 8.8), like that used for the knee, gives protection without restricting movement. It is not obvious when worn under a jacket or sweater.

Splints

Some people find comfort from a back splint applied as soon as possible after a painful haemarthrosis. Splinting allows them to relax tense muscles which have contracted to guard the affected joint from movement. As soon as pain and tenderness have abated following treatment, and perhaps aspiration with appropriate clotting factor cover, the splint should be removed and exercises started to build up muscle power and restore joint movement. The splint can be reapplied between treatments, or used at night for a week or two without affecting this return to normality. It is very important that splints are only used as short-term remedies. Splints used for more than two or three days will lead to muscle wasting and reduction in joint range. The only exception to this is immobilization which is sometimes required for the orthopaedic management of long-standing bleeds in target joints.

Immediately after a haemarthrosis, the joint will have taken up the position of maximum comfort, and the first splint should conform to this position. As function returns, the splint should be adjusted. This is called serial splinting. If this is not done a loss of joint range will follow. Final splints for knee and elbow should be straight to encourage full extension, those for the wrist and forearm should be angled to hold the hand and fingers back, and those for the ankle angled to stretch the Achilles tendon and flex the joint beyond 90°.

Short-term splints are made from plaster of Paris (POP), fibreglass, or plastic. Fibreglass splints (Baycast; Deltacast; Dynacast) set more quickly and are light and water-repellent, but they are far more expensive than POP. Both they and POP splints are easy to make and people who travel or live some distance from a treatment centre can make their own. The requirements are as follows:

1. rolls of POP or fibreglass bandage (width will depend on age: for an adult 15 cm (6 inches) for arm, 20 cm (8 inches) for leg;
2. rolls of stockinette;
3. rolls of soft padding material;
4. rolls of loose-mesh cotton bandage (Crinx) or crêpe bandage;
5. a sheet of polythene, rubber, or mackintosh;
6. warm water.

The limb is prepared by putting on the stockinette and wrapping it round with the soft padding. More padding should be used for fibreglass casting than POP because the former is both more rigid and rougher than the latter and, over a period of time, more likely to cause skin irritation. A surplus of material is left at each end to cover the ends of the splint. The polythene sheet is placed under the limb to protect the bed from plaster splashes.

To make a splint, a length of dry bandage is measured against the patient's limb (Fig. 8.9). The finished splint should extend as far as possible on each side of the affected joint and enclose at least half the circumference of the limb. For instance, a back splint for an affected knee should extend from just above the ankle to 5 cm (2 inches) below the groin. The measured length is laid on a flat surface and the thickness increased by laying further lengths on top, folding the ends down as the bandage unrolls. Ten layers should be enough for a short-term splint. If the first roll runs out start another.

The dry, layered bandage is lifted by both ends which are held as the splint is placed in warm water. Completely immerse the bandage for a few seconds. Remove and squeeze excess water out. Hold one end lower than the other. Gather the bandage against the lower hand, still keeping hold of the ends (Fig. 8.10).

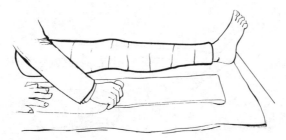

Fig. 8.9. Making a plaster back-splint—limb wrapped in padding—measuring the POP bandage.

Fig. 8.10. Making a plaster back-splint—keep hold of both ends of the POP bandage—squeeze surplus water out.

The splint is taken to the patient and, with the ends pulled out, moulded to the limb. Whilst still wet, it is wrapped round firmly with the cotton or crêpe bandage (Fig. 8.11). The ends and corners are turned over to prevent them digging into the skin after drying. The limb is held in the position of maximum comfort while the splint dries; this takes a few minutes. The splint gets warm as the chemical reaction which hardens the plaster takes place.

The damp plaster should never be held with the tips of the fingers, as indentations may occur and on hardening will dig into the patient. For a leg splint, the palm of the hand is placed under the heel to hold the leg. For an ankle, the calf is supported just below the knee, or the toes are held. For an elbow, the wrist is held.

Plaster takes at least a day to dry out fully and it is best to leave the splint in position for this time. If it needs to be used for a longer period, the cotton and padding are cut through on one side opposite the splint with a round-ended pair of scissors, and the splint gently removed. Surplus cotton and padding are trimmed, leaving about a 2

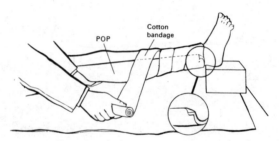

Fig. 8.11. Making a plaster back-splint—wrapping with cotton bandage—hold POP in palm not fingers—turn back corners.

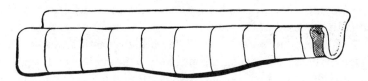

Fig. 8.12. Making a plaster back-splint—finished splint suitable for storage until needed.

cm margin which may be turned back and held with sticky tape (Fig. 8.12). The splint is reapplied and held in place with a crèpe bandage, which may be reused several times.

If the splint cracks at the joint, it may be repaired once or twice by layering several short layers of wet POP over the damage. Once a straight, firm splint has been made it is worth storing for future use.

Plastic splints

Lightweight thermoplastic materials soften in boiling water and can be moulded to the correct shape before cooling and hardening again. The thin sheets of heated material are easily cut with scissors and the splint may be remoulded to conform to the limb as a joint regains function. Splints may be braced by sticking supporting strips of the material over lines of stress. Orthoplast is the ideal material for home use, but is much more expensive than plaster of Paris.

Long-term appliances

These are rarely needed now in places where the ready availability of early treatment for bleeds has led to a decrease in chronic arthritis.

In brief, there are three main reasons for long-term appliances:

- relief from weight bearing;
- maintenance of limb posture; and
- protection.

The leg calliper is still occasionally used to take weight off a severely-disrupted knee which has been subjected to frequent bleeds. The two metal strips are attached at the bottom to the shoe heel and at the top to a rigid padded ring which encircles the thigh and contacts the pelvis. The weight of the body is transferred from the pelvis to the ring and down the strips to the heel, instead of down the leg and through the knee joint. Nowadays, callipers are very rarely used for periods of more than a few months, although they may still have a longer use in the chronically affected older patient.

More often used now are lightweight padded plastic splints which are moulded to the individual's limb and fit like the shell of an insect. These splints are less unattractive than the traditional calliper, are easy to fit, remove, and clean, are unobtrusive under clothing, and above all allow at least partial muscle action to continue during the immobilization of a limb. This cuts down the time needed for rehabilitation once the need for splinting has passed, and reduces the likelihood of further bleeds in a limb weakened by prolonged disuse. A particular example is the ankle splint which is worn next to the skin and fits inside a normal sock and shoe.

Sometimes very simple changes to footwear can be very effective. Differences in leg length, perhaps the result of arthritis in one knee, can be corrected by shoe raises. An unbalanced gait may be solved by inserting flared soles into shoes. These orthotic devices are cheap and easy to obtain. However, their use should be assessed regularly by the physiotherapist to see if they are still required and, if they are, whether they need adjusting.

A note about wheelchairs

We throw wheelchairs out of the Newcastle Haemophilia Centre! People with haemophilia are not invalids and wheelchairs are necessary *only* when an injury or surgery prevents them walking temporarily. In the very rare instance of a person with haemophilia being severely handicapped, a wheelchair will be prescribed and carefully measured to suit the individual patient. A chair of the type that folds to fit in a car boot or on a bus is recommended for most of these people. It should have pneumatic tyres, a firm but well-padded seat and side arms, and adjustable leg rests. The braking system should work from both sides so that it can be used if one arm is incapacitated. Care should be taken to ensure that a child does not use his chair more than is absolutely necessary. The usual alternative to a wheelchair for short-term use in childhood is a folding buggy.

9

Surgery and dental extractions

The decision to operate on someone with haemophilia is not an easy one for either the patient or the surgeon. Older patients, who have grown up through the years of inadequate treatment, will be afraid of uncontrollable bleeding. Parents of the younger affected boys may have heard stories of long or difficult spells in hospital following the simplest dental extraction. The surgeon will want to be sure that bleeding can be controlled until the wound has healed properly after the operation.

Fear of surgery is, of course, not confined to those with bleeding disorders. No normal person likes hospitals and operations, and the question 'Will everything go right?' is a natural and healthy reaction to the news that surgery is necessary. There are fortunate patients who can whole-heartedly put their trust in a particular doctor or a particular faith, and relax. But for most the worst part of an operation is fear of the unknown: 'What if . . . ?' It is hoped that this chapter will help people to understand and thereby lessen this anxiety.

The first thing to say is that no one contemplates an operation on someone with haemophilia unless it is absolutely necessary. In unskilled hands even the simplest operation may be dangerous, and this was recognized in ancient times when boys from families with a history of abnormal bleeding were excused ritual circumcision.

An operation will be necessary in two situations. The first is as an emergency procedure, for instance, in acute appendicitis. In this situation, everyone knows very well that if the inflamed appendix is not removed the patient will rapidly become gravely ill and may die; there is no alternative to surgery.

The second situation occurs when an operation is planned either to extend life (the patient has a cancer and will be cured with its removal), or to relieve suffering (frequent joint bleeds restrict work and leisure, and pain is not controlled by drugs). These situations involve 'cold' surgery and the operation is planned carefully over a period of time. Surgery, whether emergency or planned, may still be carried out on patients who are infected with HIV, or hepatitis B or C.

All operations, whether emergency or cold, are remarkably safe nowadays. There follows a description of a planned operation. The procedure for emergency surgery is considered towards the end of the section.

Arthur is a 35-year-old man with severe haemophilia A. He has a history of recurrent joint and muscle bleeds since childhood. He is married with two children and has a satisfying job with a local firm. For the past 18 months, his life has been made miserable by frequent intermittent bleeds into the left knee. The bleeds have continued despite intensive factor VIII replacement, prophylaxis, and physiotherapy. The joint was known to be damaged before the run of bleeds but, in spite of some limitation in movement, Arthur managed to lead an active life. He is now in almost constant pain, his sleep is disturbed, and he has difficulty in concentrating at work. He worries about the number of pain-killing drugs he is having to take to keep going, and about the effect of this illness on the family.

Arthur's doctor decides that the time has come to discuss an operation. He first talks to the orthopaedic surgeon on the haemophilia team and they look at the X-rays and scans of the joint. These show the typical changes of haemophilic arthritis or arthropathy. There is destruction of the smooth outlines of the articular cartilage and small cysts have formed at the ends of the bones. Arthur is seen at the hospital and the possibility of operation talked over with him and his wife. A final decision is then made to proceed.

From this point, the number of people involved begins to grow rapidly. A bed must be found, and a date fixed for the operation. This date must coincide with all the members of the haemophilia team being available. The haematologist alerts the technicians who will be busy carrying out the necessary tests in the coagulation laboratory. He also informs the blood transfusion service of Arthur's blood group. He calculates the amount of clotting factor replacement therapy required to cover both the operation and the post-operative period, allowing for possible complications. No operation can start until enough clotting factor cover is available. In some parts of the world, haemophilia centres may have to work together to find enough product to cover a major operation.

Arthur is admitted to hospital one or two days before the planned operation. He has been advised to stop, or at least cut down, his smoking in order to lessen the risks of a chest infection after his anaesthetic. Apart from his knee problems, he is found to be in good health. A chest X-ray is taken to confirm the absence of any factors likely to interfere with the anaesthetic. His blood pressure is taken and an electrocardiogram (ECG), which is a recording of the heartbeat, performed. His urine is checked for sugar, protein, and blood.

Arthur will be asked to sign a form consenting to the anaesthetic and the operation. The anaesthetist will call to examine him on the day before operation. Blood is taken into a series of different coloured tubes. One portion goes to the blood bank where it is used to crossmatch his blood with packs of red cells which may be needed for replacement at operation. A portion is studied to make sure that he is not anaemic. Part is used in an assessment of his kidney function in case drugs which may interfere with this are required later. Another portion goes to the coagulation laboratory where his factor VIII level is checked and, most important of all, the absence of inhibitors to factor VIII is confirmed. If inhibitors are found the operation may be cancelled at this stage.

All is well, and the day of operation arrives. Arthur has not been allowed to eat or drink since midnight. This precaution is taken because the anaesthetic may induce vomiting which would be dangerous whilst he is unconscious.

Two hours before the operation, a drip is placed into a vein on his forearm. The drip will stay up until he is well on the road to recovery. It serves as a pathway for the injection of factor VIII and any drugs that may be required, including pain killers, and for the replacement of fluids.

A large dose of factor VIII is now given. The type of product used and the size of the dose have been determined by calculating the response required for the particular operation. This is done by knowing Arthur's weight and the amount of factor VIII present in the different products available. The dose is given as fast as possible and ten minutes later a specimen of blood is on its way to the coagulation laboratory for factor assay. This assay is vital. It tells the doctor in charge that Arthur has responded to the dose and it tells him the height of the response. From this he can work out future doses more accurately. If the response is good, it acts as an additional confirmation of the absence of inhibitors. The surgeon is now told it is safe to proceed.

Shortly before operation Arthur is given a drug ordered by the anaesthetist. This is the 'pre-med', and its purpose is two-fold. It makes him feel drowsy and lessens his anxiety, and it starts to dry up the natural secretions in his mouth, secretions that might interfere with the anaesthetic. At this time too, Arthur is asked to put on a loose gown. His leg may be washed with disinfectant and wrapped in sterile towels before the porter arrives to take him to theatre.

Arthur will not see the operating theatre. The anaesthetist and the anaesthetic nurse will welcome him to a room nearby. Here a small injection into the drip or a vein, or a few deep breaths of an anaesthetic gas, will be the last he knows until he wakes up in his bed after the operation. There is nothing to fear here. Contrary to Hollywood opinion, the theatre suite is a calm, relaxed place. The surgeon will be talking about sailing or growing tomatoes while he scrubs up with his assistants and the nurse who will look after the instruments during the operation.

The factor VIII transfusion beforehand has completely corrected Arthur's haemophilia. As he lies on the operating table, he is a normal patient undergoing a normal procedure without fuss or hurry. The surgeon will handle him gently, and the operation will proceed quietly.

Arthur is undergoing an arthroplasty, the replacement of the damaged joint surfaces with new, artificial ones made of metal and plastic. Orthopaedic operations on the limbs are performed with the aid of a special tourniquet pumped up round the limb, above the site of the surgery. This makes the operating field practically bloodless and Arthur does not even require a blood transfusion. The tourniquet is removed at the end of the operation. The whole procedure takes less than two hours.

Back in the ward, he comes round slowly. His left leg will be painful and he will feel generally stiff and achy. Drugs will be given intermittently to help him overcome the pain.

Over the course of the post-operative days, Arthur will receive regular factor VIII infusions to maintain his blood level in a range which will promote healing without bleeding. Two days after the operation, the thin plastic tubes used to drain any blood from the new joint into vacuum bottles at the side of the bed are removed. The physiotherapist will urge him to keep his muscles active whilst in bed. If he is lucky, an occupational therapist will do the same for his mind. Surprisingly, he may feel depressed for a few days after all the excitement and anxiety leading up to his operation. This is normal.

The length of time in hospital will depend on the operation and the surgeon. In Arthur's case it is three weeks. The wound will be inspected when the stitches or clips are removed on the tenth day. More physiotherapy will eventually help to restore his confidence in walking. Three months after his operation he is back at work, well and happy that he took his decision. His life is no longer disrupted by frequent bleeding and, although he will still have pain for some time, it is much more tolerable than before the operation.

Emergency surgery

The procedure for emergency operations is a condensed version of the above. Because it is imperative that the operation is started quickly, difficulties in factor cover must be met as they arise. Factor responses and inhibitor screens are performed during the operation, and the search for adequate replacement therapy may involve calls for suitable blood donors and the use of products from other areas. At least in developed countries, the network of haemophilia centres, blood transfusion services, and commercial enterprises are always able to provide for the needs of any patient at any time. They must, however, be used wisely, as calls for help are very expensive and take up the valuable time of numerous people.

Orthopaedic surgery

The orthopaedic operation described earlier was an arthroplasty, the construction of a new movable joint. Arthroplasty involves the replacement of part, or even the whole of a joint with artificial parts made of metal and tough plastic—a prosthesis. The idea is not new; artificial hip joints were being fitted over 30 years ago. Although with new materials and increasing knowledge about the mechanics of the hip modifications have been made, the original concept has stood the test of time and tens of thousands of hip replacements are performed every year. Knee arthroplasty is proving equally successful in haemophilia. The type of joint prosthesis used will depend on the damage already sustained by the joint and the preference of the orthopaedic surgeon concerned.

Prostheses exist for other joints as well, but are less successful. It is often better to have an arthrodesis (see below) of the ankle for haemophilic arthritis than an arthroplasty. Artificial shoulder joints have been developed but must still be regarded as being experimental. Total elbow replacements must be regarded in the same way, although partial replacement has been used successfully in some haemophilia centres. The removal of the head of the radius, which has become enlarged as the result of recurrent bleeds, may increase the range of movement in a chronically damaged elbow joint, and relieve pain. This operation is often performed at the same time as synovectomy (see p. 200).

A person with a prosthesis is more liable than usual to get an infection in the joint, a condition called septic arthritis. This is a particular hazard of HIV infection when resistance to infection is low. Antibiotic cover is recommended to cover dental extractions and prompt treatment is needed for any septic lesions in someone with a prosthesis.

Whenever arthroplasty is contemplated, the patient must understand that the operation itself can be very difficult to perform. If the surgeon cannot achieve a good result because of the state of the bones he has to work with, he may be forced to perform an arthrodesis or joint fusion, that is a stiffening of the joint by joining the bones together. This is not as bad as it seems. An arthrodesis can relieve pain with the minimum of disruption to the patient's lifestyle. It is usually simpler than arthroplasty and is more unlikely to fail. Whereas successful arthroplasty depends on the integrity of the metal, plastic, and bone cement used, an arthrodesis is dependent only on the fusion of the bone ends that form the union. Arthrodesis is especially helpful for a painful ankle joint in adult life, because the resulting restriction in movement is hardly noticeable.

When a knee is arthrodesed, the joint itself is removed by slicing through the top of the tibia and the bottom of the femur. The cut ends of bone are then clamped or screwed together, the tissues sewn up and a plaster case applied (Fig. 9.1). With the removal of the joint the operated leg will obviously be a little shorter than the other leg. This allows the now stiff limb to be swung through the arc of walking without scuffing the foot on the ground. The surgeon will also have angled the bones slightly to allow the ball of the foot and toes to push off the ground at just the right angle when walking.

Joint replacement or stiffening are not necessarily the first operations of choice in someone with haemophilia. Sometimes realignment of long bones can bring increased stability and loss of pain to a damaged joint. The operation to do this is called an osteotomy, and is popular in some haemophilia centres.

Another operation is called synovectomy. When a joint has been subjected to recurrent bleeds it is referred to as a 'target joint'. It then becomes the target for maximum treatment in an attempt to save maximum efficiency and prevent further deterioration. Inside a target joint the synovial membrane will have grown abnormally in response to the challenge of blood breakdown products, including iron. Instead of a smooth translucent pinkish sheet of tissue, the synovium of a target joint looks like a dirty brown and rather battered sea anemone. In each of the 'anemone's fronds' are tiny blood vessels which may burst when a 'frond' is caught between the moving surfaces of the joint. When appropriate factor replacement and other conservative measures have failed to curb recurrent bleeds, the removal of the synovium may be considered. Removal of most of the membrane is straightforward, and it allows the joint and its supporting muscles time to recover before regrowth occurs. In our centre synovectomy of the elbow has been partially successful. Other centres advocate its role in the control of knee bleeds.

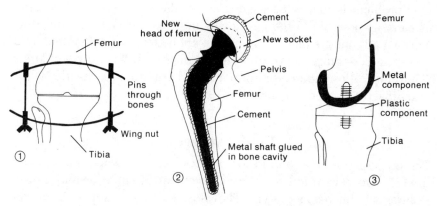

Fig. 9.1. Some examples of joint surgery. (1) Shows an arthrodesis of the knee joint, from the front. The ends of the femur and the tibia have been removed. The exposed bone is clamped tightly together by passing strong metal pins through the lower end of the femur and the upper end of the tibia; the pins are then put under tension using nut-and-bolt sets which link their ends. With time the bone unites, forming a strong, stiff joint, and the pins are removed. (2) Shows an artificial hip joint, from the front. A new socket has been embedded in the pelvic bone, and a new metal ball head swivels in it. The ball of the joint is held into the femur at the end of a long, very tough prong. Cement is used to help hold the components in place. (3) Shows an arthroplasty of the knee joint, from the side. The damaged ends of the femur and the tibia have been sculptured very exactly to fit the new components. These are held in place with plastic inserts rammed into the bone ends, and sometimes with cement.

Some centres perform synovectomy through an arthrosope rather than an incision. Haemophilia cover is needed as usual if this procedure is adopted.

Some doctors prefer to use chemicals or radioisotopes, like yttrium, rather than surgery to destroy the synovium. For instance, in the Los Angeles Orthopedic Hospital radionuclide synovectomy is performed in much the same way as joint aspiration under the appropriate clotting factor cover. Local anaesthetic is used to freeze the skin and a needle is then inserted into the joint space. A radioactive isotope (in this case phosphorus) is then injected, followed by further local anaesthetic and a steriod preparation. Most patients go home on the day of the procedure and wear a splint for two further days.

Procedures like this are not cheap. However, if the results of trials to date, which suggest a marked fall in the number of joint bleeds following the procedure, are confirmed, radionuclide synovectomy may have a place in developing countries without long-term access to concentrates.

One problem is that to be most effective the isotope should be used as early as possible in the joint being targeted by recurrent bleeding. This usually means that young people are better treated than old, and that great care must be taken not to interfere with the growth of bone. Continued follow-up to monitor potential harm from the small amount of radioactivity used is essential.

Dental surgery

The introduction of antifibrinolytic drugs like Cyklokapron or Amicar (p. 137) has revolutionized the management of some operative procedures in the haemophilias. The most striking example is dental extraction. Twenty years ago extractions usually meant several days or weeks in hospital and daily or twice daily blood product cover. Nowadays the average extractions are completed in one stage with one injection of the missing factor and several days' oral antifibrinolytic therapy, together with an antibiotic. Simple extractions can be performed as an out-patient procedure. More complex operations or the presence of inhibitors mean that a hospital admission will be necessary. Even injections into the gums (dental blocks) must never be given to someone with haemophilia unless his clotting factor deficiency has been corrected. As with any injection into the tissues, internal bleeding can be provoked with devastating results. *Never forget to tell a dentist about a bleeding disorder!*

In spite of this advance, the precautions described above still apply. Those with haemophilia who insist on surgical treatment in ill-equipped hospitals or surgeries do so at their own risk. Far better to be safe in the knowledge that inhibitors have been excluded by laboratory tests, and that full back-up facilities are available immediately should anything go wrong.

PART FOUR

Problems

10

Inhibitors

Inhibitors are not the end of the world! People who develop inhibitors to clotting factors can nowadays enjoy good and productive lives, and even major surgery is possible in special circumstances.

What are inhibitors? In coagulation they are substances which 'inhibit' the actions of clotting factors. To do this an inhibitor attacks the relevant clotting factor directly. In the process the clotting activity of the factor suffers. When inhibitor attack is strong the effect can be marked. When it is weak much of the activity may be left and the clotting process survives relatively intact.

Inhibitors are antibodies. However, the term is not synonymous because not all antibodies found in clotting are inhibitory.

Normally antibodies act as part of the body's defence mechanisms. When the body is alerted to attack by something foreign to it, it responds by producing antibodies. This is part of the immune or defence response, and we make use of it to protect people against specific disease by immunization.

When immunization against infection, for example tetanus, is given, the body's antibody defence system is put on alert. The memory of this alert is retained and used should a real attack occur. Then antibodies are produced in profusion to defeat the invader, in this case the organisms that cause tetanus. Although the memory which follows immunization is usually permanent it does weaken. This is why booster doses are required, especially if a person is at especial risk of contracting a particular disease. By convention, whilst the defender is an antibody the invader is an antigen, and the fight between them is referred to as an antigen–antibody reaction.

In some disorders antigen–antibody reactions are clearly inappropriate, the body itself undergoing damage. In these cases something has gone wrong with the defence mechanisms and an antibody response is triggered to the body's own cells; in football terms, the body 'scores an own goal'. Diseases as diverse as diabetes mellitus, pernicious anaemia, rheumatoid arthritis, and thyroid dysfunction are often the result of such an inappropriate response. Collectively they are called 'the autoimmune disorders'. In each of them there is a failure to recognize some of the body's cells as 'self' or 'friendly', as belonging to their body. These cells are attacked by antibodies and destroyed, the result being the development of an autoimmune disease. Although the underlying causes of these diseases are very complex we do know that there is sometimes a positive family history, and therefore a genetic influence at work. The same is true with haemophilic inhibitors, although, thankfully, it is rare for more than one member of a family to be affected.

In over 70 per cent of people with severe haemophilia A, the body is able to recognize factor VIII as friendly, so that infusions present no inhibitor problem. However, in about 30 per cent of severely affected people with haemophilia, the transfusion of factor VIII provokes an immune response and inhibitors are formed. These inhibitors are directed against the factor VIII, and their purpose is to destroy it. Thankfully, most of these inhibitors are weak and easily overcome. Inhibitors may arise against clotting factors other than factor VIII in people with severe deficiencies, but they are very rare. For an unknown reason far fewer people with severe haemophilia B develop an inhibitor against factor IX. In most cases of clotting factor inhibitors it is presumed that the clotting factor molecule the body is making is so strange that when a normally active molecule is transfused it cannot be recognized as 'self'. In some patients the immune system itself probably goes wrong. In both cases the result is inhibitor formation.

Factor VIII inhibitors occasionally occur in people without haemophilia, usually as part of another autoimmune disease, but sometimes in association with childbirth, malignancy, or as a side-effect of some drugs. These patients may need the same treatment as people with haemophilia, including factor VIII therapy.

High and low responders

People with haemophilia are not born with inhibitors. It requires challenge from the relevant clotting factor to provoke inhibitor response. However, it seems that a certain, small population of people with haemophilia are born destined to produce an inhibitor. There has been a good deal of research to identify this population in order to try to protect them in some way, and recent work on the factor VIII gene suggests that in the near future we may be able to predict who is at risk. We also know that inhibitor formation is unlikely to occur if it has not already done so by the time of the hundredth exposure to the clotting factor. There is, however, new evidence that some people exposed to one of the new high-purity or recombinant concentrates may develop an inhibitor despite earlier treatment.

Inhibitors are usually discovered in one of two ways. Either the patient or his family notice that treatment is becoming less effective, or the presence of an inhibitor is discovered in the course of a routine laboratory check. People undergoing any sort of surgical procedure must be screened for inhibitors (p. 197); those on home therapy should receive regular checks. The strength of an inhibitor will vary from person to person; stronger, more powerful inhibitors are referred to as 'high-titre' inhibitors. Conversely, weak inhibition occurs with 'low-titre' inhibitors. As the laboratory tests designed to measure inhibitors vary throughout the world it is difficult to give precise figures. The two most common laboratory methods used originated in Oxford, England, and Bethesda, Maryland; hence 'Oxford' and 'Bethesda' units. Very roughly, a Bethesda unit is double an Oxford unit, but caution is needed as the way of measuring differs considerably, and such a comparison may not be valid. As a guide, low-titre inhibitors are those below 10 Oxford (20 Bethesda) units. High-titre inhibitors may rise to a reading of several thousand in either system.

In deciding about the best way of helping someone with inhibitors it is not the titre which matters but the way in which it responds to an infusion of the clotting factor (Fig. 10.1). If the response is small (the most common) and the rise in titre only a few units, the person is said to be a 'low responder'. 'High responders' are people who mount a large defence against the intrusion of the clotting factor, and in whom the titre rises by many units. In scientific terms the response is called 'anamnestic', literally a response triggered through the body's immune memory. It is important to realize that a state of high responsiveness need not be permanent; variations in response may occur with time or therapy.

Treatment of people with inhibitors
The first signs

Although at present we are unable to predict inhibitor formation, a patient is sometimes seen at the precise time that his body is mounting its first antibody defence. There is some evidence that special treatment at this time will block the response and perhaps trick the body into accepting the clotting factor. Although the evidence is slim, some doctors think such treatment is worth trying. It involves giving medicines designed to damp the process of immunity.

These immunosuppressive drugs include steroids, cyclophosphamide, azathioprine, 6-mercaptopurine, and cyclosporin. All these drugs have established places in the treatment of people with cancer or transplanted organs so their side-effects are known. It follows that the most careful surveillance must be continued if they are used in haemophilia.

The low responder

Here treatment is often very simple. Most low responders can be treated in exactly the same way as people without any inhibitors. Indeed, in some people who are shown by laboratory tests to have developed low-titre inhibitors, intermittent treatment or prophylaxis with low-dose factor replacement results in a disappearance of the inhibitor; in these cases it is assumed that the body has been forced to learn to accept the factor as 'self'. Bleeds are treated in the normal way as soon as possible after recognition. Occasionally an increased

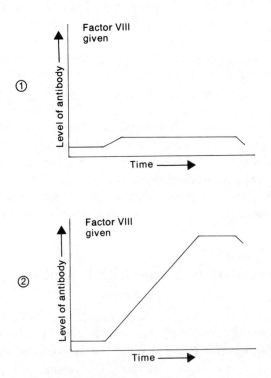

Fig. 10.1. Low and high responders. In (1) there is a 'low response' to factor VIII, hardly any rise in antibody occurring. (2) Shows the graph of a high antibody responder following challenge with factor VIII.

dose of factor is needed. Only careful surveillance over a period of time can suggest to a patient and his doctor that dosage should be altered.

The high responder

There is no single answer to the management of the high responder. Several methods of treatment have been tried. None have been entirely successful in every case. In haemophilia A the methods are described below.

Treatment with low-dose factor VIII (either on demand or as prophylaxis)

This is unlikely to work as well as in the low responders but is worth a try initially, provided careful laboratory surveillance is maintained.

Treatment with high-dose or ultra high-dose factor VIII

Occasionally a simple adjustment of factor VIII dosage to a high level will abort most bleeds without causing an undue rise in inhibitor level. Obviously the earlier the dose is given the better. Prophylaxis with this method is unlikely to be of benefit. Ultra-high dosage was pioneered in the Bonn Haemophilia Centre in Germany. Just as regular treatment with low-dose factor VIII may result in a loss of low-titre inhibitor in some people, massive doses swamp the body's immune resistance in some high responders. The regime used in Bonn involves the infusion of well over 300 000 units of factor VIII a year and, as the usual annual dosage per patient in developed centres is around 50 000 units for the treatment of haemophilia A, the method is too expensive for many Centres. The Bonn method has been tried by other doctors with variable results. It is very, very expensive but it does work for some patients.

Treatment with animal factor VIII

To some readers this may seem a novel approach, yet animal factor VIII formed the mainstay of surgical treatment 30 years ago. In those days there was insufficient human product to check post-operative bleeding and factor VIII from pigs or cows was used instead. These early products were very effective, but they had two disadvantages. Firstly, they caused the platelets in the blood of the recipients to clump together, producing a lack of platelets (thrombocytopenia). Secondly, they ceased to be effective after some 8–10 days of treatment. The effect on platelets was more marked with the bovine (cow) product, and this was usually held in reserve until the pig (porcine) product had started to lose its efficacy.

More recently porcine factor VIII has been developed to produce a product (Hyate:C, from Porton Products, United Kingdom) without the platelet clumping effect. This has proved to be very useful in some patients with inhibitors. Once again any treatment should be under close laboratory surveillance, as inhibitors against human factor VIII can also have an adverse effect on the animal product and vice versa.

The bypassing of factor VIII

If factor VIII and its inhibitors could be bypassed by using another clotting pathway, bleeding should stop. Such an alternative is provided by the extrinsic pathway (p. 49) which is triggered by injury and the release of tissue factor. The clotting factors involved in this pathway are VII, IX, and X and they have been used for the treatment of people with high-titre inhibitors. They appear to work best if steps are taken during manufacture to convert them from inert proteins to active enzymes. In their activated form (denoted by the letter 'a' after the factor number) they are thought to act immediately after injection by driving the final common pathway to the formation of a clot.

Obviously, there could be some danger in this because if the process is driven too hard or too fast excessive clotting may lead to thrombosis. For this reason antifibrinolytic drugs like Cyklokapron are usually avoided when bypass therapy is being given.

Treatment with factor VIIa

Factor VII is one of the proteins known to work early in the process which leads to the formation of a blood clot. It starts to do this once it has been converted from its original, inactive form (VII) to an active form called an enzyme, VIIa, where the 'a' stands for 'activated'. It is tissue factor, the chemical soup released from injured cells, that activates VII in this way. Together with factors IX and X, factor VII then forms a complex which initiates clotting.

Two forms of pure VIIa are now available, one made from human plasma, and the other by genetic engineering produced by Novo-Nordisk of Denmark. The plasma product has been shown to be effective in many closed bleeds in a French study of factor VIII antibody patients, and the recombinant product (rFVIIa Novo) also looks very promising, and could help people with factor IX inhibitors as well.

Treatment with factor IX

Factor IX bypass therapy has been in clinical use for 20 years. The older factor IX concentrates also contain factors II, X, and sometimes

VII as well, so it may not simply be the factor IX which is effective in bypassing factor VIII inhibitors. That these products are effective has been shown in at least two studies of inhibitor patients. They do not always work, but they work often enough to provide a back-up form of therapy if other products fail.

The factor IX products have been divided into two groups, 'hot' (or 'activated') and 'cold' (or 'inactivated') products. The former, marketed as FEIBA (Factor Eight Inhibitor Bypassing Agent), and Autoplex, contain clotting factors which have been treated during manufacture to trigger clotting as soon as they are introduced into the body; in other words, the process of coagulation is already underway before injection. The latter, cold products have received no special modification during manufacture, and are products previously used routinely for the management of people with haemophilia B. Although the studies referred to above have been with hot products, considerable debate continues about their value in comparison to the ordinary cold factor IX concentrates. Why?

Firstly, there is no infallible method for measuring the activation of a clotting factor. In other words, there is no way of saying: 'If I put 10 litres of fuel in this car it will go x miles'. Therefore there can be no laboratory comparison between products and no scientific measure of their relative efficacy.

Secondly, most studies have been on closed bleeds rather than on blood loss that can be measured directly.

Thirdly, at least a quarter of the bleeds studied stopped with placebo (inert) treatment when it was used instead of the clotting factor concentrate.

But most telling of all, in practical terms, is the finding that cold or inactivated products seem to work equally well in stopping bleeds in many people with inhibitors. Because the hot or activated products are designed specifically to induce clotting in the body they must carry the inherent danger of overdoing things. Clots have been formed in places other than those intended both in people with haemophilia B and in haemophilia A inhibitor patients so, once again, anyone contemplating treatment with these products should be under the careful surveillance of a knowledgeable doctor.

The use of localized clotting agents

It is not always necessary to have a clotting factor circulating in the bloodstream to control bleeding, and there are a number of tricks that can be used locally in certain circumstances. Troublesome nosebleeds may respond to the application of drops containing Cyklokapron. We use the intravenous preparation drawn up into a small syringe. A couple of drops into the nose followed by pressure over the offending nostril, repeated if necessary, are worth a trial.

In America some patients have found that the insertion of a small piece of highly salted pork stops bleeds; the meat falls out after a day or two. When we heard of this seemingly idiosyncratic method we tried a salt gel and found that it did help a few people. Presumably the salt causes small blood vessels to contract down, stemming the bleeding. Russell's viper venom is no longer available in the United Kingdom as a local haemostat but topical thrombin is, and its application to surface wounds in a pledget of absorbable material, backed by a non-stick dressing and followed by firm pressure, often works. Larger wounds, including bleeding tooth sockets, respond to plugging with pastes made of clotting factor concentrate mixed with a little diluent; if pressure is applied as well so much the better. An antifibrinolytic (p. 137) should be given for several days to stop the resulting clot breaking down before wound healing is well advanced. In addition, dental sockets and large wounds should be protected from infection by antibiotic cover for at least one week. When possible, wounds should also be protected by dressings or splints and immobilized as far as possible. Cyklokapron mouth washes and fibrin glue have also been used to present bleeding.

Bleeding into joints may set up the inflammatory response called synovitis, which may in turn lead to further bleeding. When synovitis is diagnosed it may be controlled with the careful use of a short course of steroids. These anti-inflammatory drugs have many side-effects and should not be used over long periods, especially in growing children, without good reason and expert surveillance. When they are prescribed in haemophilia, enteric-coated preparations should be used and they should only be taken after meals. The special coat allows the medicine to pass through the stomach into the bowel, thus preventing irritation of, and perhaps bleeding from, the stomach lining. Preparations used are called prednisolone (UK) or prednisone (USA). A soluble product is available for children.

Surgery in someone with inhibitors

It really goes without saying that all the tests that should precede any surgical procedure in someone with haemophilia must be performed when inhibitors are involved. If planned surgery cannot be avoided it can go ahead with few additional risks by using available technology. The various ways by which an inhibitor might be overcome have already been listed; cover for surgery might involve one or all of these.

Table 10.1 Summary of available treatment in the management of clotting factor inhibitors

1. Immunosuppression*	May work if given very early
2. Usual factor replacement*	Works in most low responders
3. High-dose factor replacement*	Works in remaining low responders. May work in high responders
4. Ultra high-dose factor replacement	May induce tolerance in high responders. Very expensive
5. Porcine factor VIII	Works when inhibitors to porcine VIII low
6. Factor VIIa*	Promising early results from clinical trials
7. Factor IX ('cold')	Well worth a try if factor VIII is not effective
8. Factor IX ('hot')	Worth a try if 'cold' IX is not effective
9. Plasmapheresis*	Useful back-up procedure for surgery or during major bleed
10. Local agents*	Effective in certain sites
11. Steroids*	Dampen joint inflammation. Should only be given short-term, especially in children

*Applicable to factor IX as well a factor VIII inhibitors

Any planned operation is arranged to take place when the level of inhibitor in the patient's blood is as low as possible. If factor VIII is being used, this may mean withdrawing everyday treatment of bleeds for a while and plasmapheresis using a special column designed to remove the inhibitor as blood passes through it might be needed. Once the inhibitor level has fallen and the patient has been prepared for surgery in the routine way (p. 197), either human or porcine factor VIII is given, just before or during the operation. Clotting occurs in the wound and the clot is prevented from breaking down by antifibrinolytic blockade with the provision that antifibrinolytic drugs must nowadays be used with caution in anyone already transfused with a bypass agent, or factor IX concentrate. Local measures of securing the clot may be used. Some surgeons use a laser beam to cut tissues instead of a scalpel. The laser clots as it cuts, so loss at this stage is minimal.

It usually takes 7–10 days for an inhibitor to rise after factor VIII challenge and about 2 weeks for the peak value to be reached. In this time, repeated infusions or a continuous infusion of factor VIII can be used to protect the wound. If bleeding occurs, back-up protection can be provided by using one of the bypass factor IX agents. Plasmapheresis may also be used to lower the inhibitor level again.

Very major surgery has been performed successfully using these methods. Obviously it helps if there is time for meticulous preparation, but similar methods may be used in the event of emergency, for instance appendicitis or after a road traffic accident. Because inhibitors to porcine factor VIII are less frequent than human, the animal product has a special place here.

Teeth may be extracted from someone with high-titre inhibitors without the need for prolonged factor treatment. When factor VIII is to be used it is transfused as the extractions take place, after intravenous injection of an antifibrinolytic. Because of the danger of thrombosis antifibrinolytics are not prescribed when factor IX or another bypass agent is being given. The sockets are plugged in the way described above and are protected with a specially moulded plastic thermolabile splint which drops off by itself when healing is complete. Treatment of inhibitors to clotting factors other than factor VIII is by plasmapheresis, and high-dose or continuous therapy. The only difference is that there are none of the bypass agents which seem to be useful when factor VIII inhibitors are involved.

Way of life with inhibitors

It is remarkable to see how fit and well both children and adults with even very high-inhibitor titres can be. Obviously there is more of a threat to them in the event of serious injury, but people with inhibitors can live normal everyday lives, and participate in most of the same exercises and activities as their peers. The question of restriction has been discussed elsewhere (p. 23). Naturally, thoughts of restriction, of 'living within the bounds of one's disorder', are more likely to arise when an inhibitor is present. But parents should not allow the presence of inhibitors in their child to cloud his life. In a way it is even more important than with a non-inhibitor boy to teach him to look after himself and to grow up with strong muscles and joints. Low responders should be treated in exactly the same way as those without inhibitors. The lifestyle of high responders should be discussed more fully with an experienced doctor, but when everything has been considered, the real choice may lie between the relative importance of quality and quantity of life.

11

Side-effects of treatment, 1

All medicines have more than one effect. The primary effect—the reason we take the medicine in the first place—is usually sufficiently beneficial to outweigh the other results or side-effects, but this may not always be so. Aspirin may well cure a headache but at the same time it is reducing platelet stickiness, and it may provoke bleeding. Codeine dulls pain but at the same time it reduces the motility of the bowel, causing constipation. Often these secondary effects are dose dependent and easily relieved, but the more powerful the drug the more serious the other effects are likely to be. Every time a doctor prescribes a drug he weighs up the potential benefits against the potential costs to his patient. If the benefit could be life, for instance in cancer chemotherapy, even drugs with very marked and potentially lethal side-effects are worthwhile.

The treatment of haemophilia is no exception, and this chapter explains the known side-effects of blood product transfusion. Before describing them it should be emphasized that behaviour is not transmitted by transfusion. All the publicity given to donors who were homosexual or addicted to drugs in the early days of the AIDS epidemic made some people fearful that they or their children could 'catch' one of these characteristics through treatment. Like so many fears engendered by AIDS this particular worry is groundless.

The side-effects of the products used to treat haemophilia may either be short or long term.

Benefit   Cost

Fig. 11.1. Weighing benefit to patient against cost to patient.

Short-term side-effects

All the blood products described in Chapter 6 must be injected intra-venously (into a vein) so that they enter the bloodstream directly. The technique of placing a needle in the vein is called venepuncture.

With modern transfusion equipment the procedure is virtually pain-less, but occasionally minor side-effects of the injection are experi-enced. These seldom amount to more than a little discomfort and soon disappear; they should, however, be reported to your doctor. Most people feel the fluid, which is usually colder than their body temperature, flowing into the vein. Sometimes the skin over the vein becomes a little reddened for a while. Occasionally a rise in temperature or a fleeting rash may be experienced, and very rarely shortness of breath, headache, or backache occur. With some products, numbness or a tingling sensation is felt in the lips, and there may be a metallic or salty taste in the mouth. These immediate effects are frequently associated with the rate of injection, and can be reduced by giving the product more slowly. People who continue to experience immediate allergic-type side-effects may be prescribed an antihistamine to give intravenously. Very rarely an allergic reaction can be severe enough to suggest that the type of product being used should be changed.

Two other rare events may complicate treatment. The first occurs when a product contains substances foreign to the patient. The result of this incompatibility is the production of defensive antibodies and the destruction of red cells, a condition called haemolysis. Haemolysis is most likely to occur after massive transfusions of clotting factor con-centrates when it may result in anaemia. It can be avoided in the small number of susceptible patients by using group specific blood products.

The second rare side-effect is thrombosis. This is associated with some of the factor IX concentrates used for the treatment of haemophilia B (Christmas disease). Very occasionally transfusion of one of these concentrates results in inadvertent clotting. No one is sure why this should be, although the most likely explanation is that one of the factors in the concentrate is activated in some way before or during transfusion. Heat treatment may increase the likelihood of this happening, and some concentrates now contain the anticoagulant heparin to counteract any thrombotic tendency. The chances of thrombosis occurring can be reduced by injecting treatment slowly, by only using freshly prepared concentrate, and by taking great care of the vein used for treatment. If there is any sign of inflammation at a particular site, another vein should be used. Because they stop the natural breakdown of clot, antifibrinolytic drugs like Cyklokapron or Amicar (p. 137) should not be used at the same time as factor IX concentrate.

The new high-purity factor IX concentrates may reduce the risk of thrombosis. They are recommended especially for surgery when the patient is likely to be bed-bound for a while and when the risk of thrombosis is higher than usual anyway. As an additional precaution some of them contain small amounts of heparin.

Long-term side-effects
Hepatitis

One of the most important side-effects of blood transfusion is hepatitis, or inflammation of the liver. The liver becomes inflamed when it is infected by one of a number of viruses. If the infection is marked enough, jaundice may result. The liver is unable to remove a naturally-occurring pigment from the bloodstream, and the build-up of pigment stains the skin and the whites of the eyes yellow. The urine becomes darker and, because pigment is no longer being excreted from the liver into the bile, the stools become pale. We know that many liver infections are not severe enough to result in the appearance of jaundice. They show themselves as mild, transient periods of feeling unwell, or only as changes in liver function measured in the laboratory, a condition that has been called 'transaminitis'. One of the reasons for following up people with haemophilia carefully is to monitor these changes.

Jaundice does not always mean that the liver has become infected. It occurs when red blood cells break down too rapidly, and when the passage of bile from liver to bowel is blocked. Usually the underlying disorder can be determined fairly easily by examining the blood and urine, but more sophisticated tests may sometimes be necessary. These include ultrasound, computerized tomography (CT), magnetic resonance imaging (MRI), or radioisotope scans of the liver and the spleen, X-rays which show up the bile passages, and liver biopsy. This last test involves taking a minute piece of liver tissue to examine under the microscope. The procedure is done under local anaesthetic, using a special needle. The same precautions that are used for any surgical procedure in haemophilia must be observed. Factor replacement is given before and for at least three days after biopsy.

A number of types of viral hepatitis are known. Those of particular interest to someone with haemophilia are called hepatitis A, hepatitis B, and hepatitis C.

Hepatitis A

Hepatitis A is a common short-lived illness. The hepatitis A virus (HAV) is spread by poor hygiene, especially by faecal contamination of food or drinking water. Infection may not be apparent; most people only feel off-colour for a while. Until recently it was thought that hepatitis A was not transmitted by blood products. However, in 1991–2 four separate populations of people with haemophilia living in Belgium, Germany, Ireland, and Italy went down with the infection. They had in common the use of one factor VIII concentrate. Despite the fact that this had been treated by a solvent–detergent method (p. 227) to remove viruses, it appears to have transmitted hepatitis A. As a consequence the concentrate was withdrawn from the market. The story is not a simple one because in other countries using the product people with haemophilia were not infected. Nor were those using similar products manufactured with the same antiviral techniques.

These outbreaks were of relevance to everyone involved in the production of safe concentrates. They suggested that the currently available methods of removing viruses, including the use of solvents and detergents, might not be entirely foolproof. For this reason manufactures began to introduce more than one antiviral technique into the process. As an example, a heating step has now been added to the solvent detergent step in the production of the concentrate thought to have previously transmitted hepatitis A.

Since the outbreaks a vaccine for hepatitis A has become available. Called Havrix, it is made by SmithKline Beecham. Three intramuscular injections into the upper arm are recommended, the second one a month after the first with a booster at one year. Everyone likely to be exposed to a clotting factor concentrate who does not already have a hepatitis A antibody should now be offered vaccination.

Hepatitis B

The hepatitis B virus (HBV) causes one of the most widespread infections known to man; an estimated 200 million of the world's population carry the disease. It is important for someone with haemophilia to appreciate this because occasionally the finding of the B viral markers in a patient results in alarm and despondency. You are not alone!

When a virus attacks the body it is recognized as an invader or antigen, and the recipient reacts by producing antibody against it. The antigen now called hepatitis B surface antigen (HBsAg for short) was originally discovered in 1965 in the blood of an Australian aborigine, and named the 'Australia antigen'. Work on the virus responsible revealed more about its structure, how it affects liver cells, and how it can be controlled. The understanding of hepatitis B is one of the success stories of contemporary medicine.

As the term HBsAg implies, the first antigenic or 'recognition' points of the virus were found on the surface. Beneath lies a core, and it was not long before markers were found which identified this structure too; they are called hepatitis B core antigen, or HBcAg. Within the core was found yet another marker which was called the 'e antigen' or HBeAg, and it is now known that this particular antigen is associated with infectivity.

Thus, the original Australia antigen has been taken apart by the researchers and has revealed itself to be like a nest of Russian dolls. Each of the antigens found has an associated antibody. Hence we have anti-HBs (or HBsAb), anti-HBc, and anti-HBe. Their recognition, together with the liver function tests, allows the course of the hepatitis B attack to be plotted, and gives an indication of when vaccination is needed to protect contacts.

Whenever someone with haemophilia is shown to have experienced hepatitis B, the information is passed to the blood product manufacturer so that identification of the infected donor can be attempted. This time-consuming and difficult job depends entirely on knowing the precise details of the blood product used within the past six months (the incubation period for hepatitis B). This is why the recording of batch or lot numbers every time a blood product is given is a vital part of haemophilia care, wherever the transfusion takes place.

Vaccination against hepatitis B is with the genetically engineered vaccine, Engerix B made by SmithKline Beecham. It is given by intra-muscular injection into the deltoid muscle of the upper arm (because uptake is very variable when injections are given into the bottom!). After the first injection, two further injections at one month and six months are required to provide immunity from hepatitis B, with boosters about five-yearly thereafter. People with lowered immunity (for instance those with HIV infection) may need more frequent injections.

In some parts of the world an older, equally effective vaccine made from human plasma may still be in use. This vaccine is safe and is given in the same way as Engerix B.

Engerix B is an example of *active* immunization. *Passive* immuniza-tion against hepatitis B is also available for those people who have not been vaccinated but who have had an injury exposing them to the risk of infection. For example, if someone stabs himself with a dirty needle whilst preparing a blood product to treat a friend or relative, specific immunoglobulin (hepatitis B virus immunoglobulin; HBIG) should be given. This product is rich in antibodies which soak up any hepatitis B virus which has been injected. HBIG is also indicated to protect the newborn babies of mothers known to carry hepatitis B infection. Passive immunity only lasts a few weeks so anyone at repeated risk should also be actively immunized at the same time.

Despite modern testing and antiviral techniques blood and blood products still occasionally transmit hepatitis B. If they prove not to be immune already on laboratory testing, the following people should be vaccinated to protect them from infection:

---

1. People with haemophilia or a related disorder who have not yet been exposed to blood products. In most cases this means that all affected children will need vaccinating. Some adults diagnosed later in life to have a moderate or mild bleeding disorder should also consider vaccination, despite the fact that their treatment of choice may be desmopressin (DDAVP). This is not because there is any risk from desmopressin, but because clotting factor replacement may be required in addition at some stage in their future.
2. People with haemophilia or a related disorder who, despite exposure to blood products, have not become immune to hepatitis B.
3. Those responsible for giving blood or blood products to people with haemophilia. This includes relatives or friends as well as hospital staff.
4. The close family contacts of someone shown to be infectious for hepatitis B. This means people with acute infection and those shown by laboratory testing to be carriers of the virus.
5. Sexual partners of someone shown to be infectious for hepatitis B.
6. Babies born to mothers who have acute hepatitis B during pregnancy, or are known to be carriers of the virus.

The delta agent The delta agent, also known as hepatitis D, is a defective virus incapable of causing damage on its own. It can only work with the hepatitis B virus when it can add to the overall effect of the illness in some patients. People protected against hepatitis B are also protected from the delta agent.

Hepatitis C

The hepatitis C virus (HCV) was discovered in 1989. Before its discovery it had long been assumed that liver disease was being caused by infectious agents other than hepatitis A or B. For this reason doctors referred to non-A non-B (NANB) hepatitis. Regular checks of the liver function of people with haemophilia had frequently shown intermittent abnormality. The recurrent peaks in the results of two liver function tests had become known as 'NANB transaminitis'.

We now know that most, if not all, this liver disease is caused by hepatitis C infection. We also know that the majority of people who needed blood product therapy before antiviral measures were introduced in the mid 1980s were infected with the virus. The test for the virus currently in use measures an antibody to hepatitis C. This antibody does not protect and a positive result indicates that active infection is present. A vaccine is not yet available.

Hepatitis C appears to be transmitted primarily by blood, either through transfusion or by sharing needles as in drug addiction. It is not transmitted to close family or other contacts, like hepatitis B. There is thought to be very little risk of heterosexual transmission. Because of this current advice is that no special measures need to be advised in the case of stable sexual relationships, and that condoms are unnecessary. Partners may wish reassurance that they are not infected by having a blood test for hepatitis C. In the rare event of the test being positive they should be followed up in the same way as the person with haemophilia (see below). If the test is negative and their sexual relationship is a long-standing one there is nothing to worry about.

In the case of those who know that they are infected with hepatitis C and sleep around, the advice is that partners should be protected by condom use. There is no difference here in the advice already given with respect to AIDS.

Since the discovery of the virus and a reliable test for its presence all blood donations have been screened. In the United Kingdom this testing was introduced in September 1991.

More than half of those infected with hepatitis C will remain well and not be troubled by liver disease. However, current knowledge suggests that around 40 per cent of infected people will continue to have abnormal liver function tests. Most of them will continue to feel well, but some will go on to develop chronic liver disease.

Chronic liver disease

It is rare for someone with haemophilia to become very ill as a result of hepatitis itself, whatever the cause. Usually the patient with hepatitis will be ill for a while, will feel nauseated and off his food (and cigarettes if he is still foolish enough to smoke them), and be a bit depressed for some weeks after the attack. He should be nursed at home rather than in hospital, and should refrain from drinking alcohol for at least six months because of its effect on the liver. His family doctor will give advice about hygiene within the home.

Whenever anybody with haemophilia thinks they might have jaundice the staff at their haemophilia centre must be told.

The worry about infection with hepatitis B or hepatitis C is that chronic liver disease might result. There is as yet no way of predicting which people are at risk of this, other than by asking them to attend for regular follow-up. At these consultations examination may reveal signs of liver disease, and blood is taken for liver function tests. If these are persistently abnormal further investigation might be suggested. This may include liver biopsy, which provides the most objective evidence of the nature and extent of any damage.

Changes in the liver are grouped under the headings:

- chronic persistent hepatitis
- chronic active hepatitis
- cirrhosis.

Most people with haemophilia who also have liver disease have chronic persistent hepatitis. For them the outlook is excellent and no treatment is indicated.

Chronic active hepatitis can lead to progressive changes and eventually to cirrhosis. This progression is slow and takes many years. Once cirrhosis, which is a chronic scarring of liver tissue, has developed a liver cancer may appear in a very few people.

Interferon

Thankfully, it looks as though many of these patients with chronic active hepatitis can be helped to recovery by injections of a drug called interferon alfa. This drug has been shown to be effective in hepatitis B infection and is currently under trial for hepatitis C.

The interferons are naturally occurring substances with roles in the body defence systems. As such they have many possible uses in the treatment of disease. Within the body interferon is produced naturally in response to infection. One of its effects is to make the victim feel off-colour with a rise in temperature and aching. That is why 'flu' feels like 'flu'! Not unexpectedly interferon alfa given as a drug has the same effect. Luckily the remedy is the same—a couple of paracetamol tablets relieve the symptoms. Some people take them half an hour before treatment. If high temperatures follow treatment and result in sweating plenty of fluid should be drunk.

Naturally occurring interferon was discovered over 30 years ago. Nowadays the drug used for treatment is made by genetic engineering. It is injected through a very small needle and can be given at home by the patient himself, his relatives, or a nurse from the family practice.

In chronic hepatitis B infection between 2.5 and 5 million international units of interferon alfa are prescribed for every square metre of the patient's body surface. The drug is given three times a week in the form of subcutaneous injections, that is injections just beneath the skin of the abdominal wall or upper thigh. A course of treatment lasts between four and six months. The dose is adjusted according to response.

In hepatitis C infection the dose of interferon alfa is 3 million units initially for a six-month period. Lower doses may then be sufficient to halt progression of the infection. As with the treatment of hepatitis B the drug must be given by subcutaneous injection three times a week. Preliminary results suggest that at least half those with progressive liver disease as a result of hepatitis C infection show a good response to interferon alfa. No one yet knows for how long such treatment should continue.

Cirrhosis of the liver

In this condition normal, fleshy liver tissue becomes scarred and unable to function properly. Cirrhosis results from many different disorders, including viral hepatitis. Probably the best known cause is long-standing alcohol abuse. No single treatment is available for people with cirrhosis, although several ideas are being explored in an attempt to switch off the process that results in scar, or fibrous tissue formation. The main hope lies in preventing the disease in the first place. The technique of liver transplantation is now appropriate for a minority of patients in end-stage liver failure. Ironically, when this very major operation is performed in haemophilia it cures the disorder because one of the functions of the new liver is to produce active factor VIII and IX.

Alcohol and the liver

Although not a side-effect of treatment, it seems sensible to conclude this account of liver disease with a word about the effect of alcohol. Alcohol is removed from the blood by the liver and if too much is drunk the liver becomes overloaded and unable to work efficiently. Repeated overloading eventually causes the irreversible damage called cirrhosis. Alcoholic liver disease and other alcohol-related illnesses and accidents are the major killers in our society after heart disease and cancer.

Alcohol is of especial importance to older, severely affected people with haemophilia because their livers have already been put under strain by repeated exposure to hepatitis viruses. Abuse of alcohol simply adds to this strain and probably makes long-term liver failure more likely. Although the newer clotting factor concentrates are undoubtedly safer, younger people who have been treated only with these products are still advised to be very wary of drinking too much alcohol. It is silly to add other potential disabilities to severe haemophilia, and the relationship between alcohol abuse and accidents can easily lead to severe bleeding, the need for more treatment, and even hospitalization.

Having said that, as a doctor concerned with the many difficulties of haemophilia, including the joint pain experienced by the older generations of patients, I am loath to suggest the withdrawal of one of life's comforts! Certainly anyone with proven, persistent liver disease should try to abstain from alcohol, as should anyone during and immediately after an episode of hepatitis. It is customary to suggest that after an attack of jaundice alcohol should be banned for six months, but the evidence to support this view is not absolute. My view is that, if alcohol in moderation makes life tolerable for someone with severe haemophilia, it should not be condemned.

As a guide to moderation people should try to abide by the recommended limits for alcohol consumption in which a unit of alcohol equals either:

- a small sherry
- a small measure of spirits
- a glass of wine
- a half a pint of beer, or
- a half a pint of lager.

In general, men should drink no more than 21 units over a week, and women 14 units. Women are advised to drink less both because they are usually smaller and because their body structure makes them more sensitive to alcohol than men of equivalent size.

Finally, remember that in pregnancy alcohol readily crosses the placenta from the mother's blood into that of her baby. Here it causes damage to delicate growing tissue and this can result in a child with mental retardation and physical deformities. It is best not to drink alcohol at all during pregnancy.

Other long-term side-effects

The most obvious of these is infection with the human immunodeficiency virus (HIV) which is discussed in Chapter 12. Viruses other than HIV and those causing hepatitis are also known to contaminate human blood. Thankfully most are removed nowadays during the manufacture of concentrates, and none are known to result in long-term disease. One of there is called parvovirus. It has a tough protein coat which protects it from solvents and detergents, and heat. A vaccine against parvovirus should soon be available.

One other side-effect, which may occur in people exposed to large amounts of clotting factor concentrate over long periods of time, is a reduced ability to respond to challenge by infection. This lowering of immunity is very difficult to measure consistently, and all that can really be said at this stage is that some patients, who are clinically very well, show abnormal results in laboratory tests. When one considers the amount of protein in addition to factor VIII or IX to which severely affected people were exposed every time they had an injection of concentrate, it is hardly surprising that their immune systems should occasionally show evidence of becoming overstressed. This topic is discussed further in Chapter 6.

Reducing the risk of harmful side-effects

There are several ways by which blood products can be made safer. None can give guaranteed safety on its own, so all manufacturers now use a combination of methods in order to ensure that any risk remaining is negligible. The methods include attention to the selection of blood donors, testing of individual donations of blood or plasma, and alterations to manufacturing procedures in order to remove viruses.

The selection of donors

Although it is generally agreed that blood or plasma should only be collected from volunteer donors, this ideal has been impossible for most countries to achieve. The costs of establishing and running a modern transfusion service capable of responding in a flexible way to the demands of all patients in need of blood products, including people with haemophilia, are enormous. As a result, it has usually been easier to buy some products from commercial companies already established in the field. The only way these companies can obtain enough source plasma to try and satisfy the demand for concentrates is to pay donors. Even with this inducement, only some 20 per cent of the world-wide demand for factor VIII can be met, and the economics of the market mean that a universal reliance on volunteer donor blood is unlikely to be realised.

Unpaid volunteers are best because they are less likely than paid donors to conceal a medical or social history of possible infection with one of the diseases transmitted by blood transfusion. However, both volunteer and paid donor blood is subjected to the following checks:

1. *A history of donor health*

   (a) *AIDS*: these people must NOT donate:

- men and women who either think they may be or know they are infected with the human immunodeficiency virus (HIV) which causes AIDS;
- men and women with AIDS;
- men and women who have ever injected themselves with drugs;
- men who have ever had sex with another man;
- men and women who have had sex with men or women living in African countries, except Morocco, Algeria, Tunisia, Libya, or Egypt;
- men and women who are prostitutes;
- in addition, of course, the sexual partners of people with haemophilia must not donate.

These exclusions are those presently in force in the United Kingdom. As HIV infection spreads they will need modification. **It is extremely important that people do not attend donor sessions in order to be tested for HIV or AIDS.**

(b) *Other conditions*: all donors should be in good health and free from any history of hepatitis, malaria, or syphilis. There are a number of other infections and medical conditions, including the taking of medicines, which may make donation at a particular time hazardous. People are given a list which they must read and sign before donating.

2. *The physical appearance of donors*
   Although it is not practicable to examine all donors physically, the doctors and attendants at a donor session will be on the lookout for any evidence of ill health.

3. *The testing of individual donations*
   Each donation of blood or plasma is tested for hepatitis B, C, HIV, and syphilis. In addition, some agencies check liver function which, when abnormal, could reveal other hepatitis viruses. This is called 'surrogate testing'.
   Other testing is performed to establish blood groups, and safety from contamination throughout collection and manufacturers.

4. *Antiviral measures*
   Even before the appearance of HIV infection and AIDS manufactures had tried to remove viruses from blood in a number of ways, none of which was very successful. The methods used included exposure to ultraviolet light, mixing with chemicals like B-propiolactone, and changes in temperature. A major problem

with these, and with present methods of manufacture, is that the harsher the process intended to remove viruses, the lower the yield of the clotting factor in the final product. This means that more donations of plasma are needed to produce an equal amount of factor VIII or IX. As well as increasing the risk of initial contamination, this means rising costs in both the collection and manufacture of concentrates, and ultimately less concentrate to go round.

So necessary is it to remove HIV from the concentrates that these considerations have taken second place, and new methods of viral inactivation are now being used despite yields which, in comparison to earlier products, are poor. The methods include:

- *Heat.* This is used with the concentrate either in a dry state as a powder, or in a wet state. Early attempts at dry heating failed because the temperature was not high enough for a long enough time to kill all the virus. However, when the dry concentrate is heated in its final bottle for 72 hours at 80 °C complete viral destruction is achieved.

  The reasoning for heating products in the wet (either as solutions or slurries) is that a high temperature can be spread more evenly throughout the concentrate, thus eliminating odd 'pockets' which might continue to harbour live virus.

  Some heated products continue to very occasionally transmit hepatitis, but there is no evidence that any continue to transmit HIV.

- *Solvent–detergent.* HIV and most other viruses which contaminate concentrates have fat in their membranes. Solvents and detergents break fat down (witness washing-up liquid) and therefore destroy fatty coated organisms. The detergents used in the manufacture of these products occur naturally in the body so there are no worries about potentially harmful effects. The solvent, which is called tri-n-butyl phosphate (TNBP), is not a natural product and after its introduction must be removed again before the manufacturing process is complete. Results to date are very encouraging with no evidence of either HIV or hepatitis transmission, other than a question about hepatitis A, which does not have a fatty coat.

- *Monoclonal antibodies.* These are structures which act like personalized magnets, and can be tailored to attract either the von Willebrand, or the clotting part of the factor VIII molecule. As liquid concentrate passes down columns containing the antibodies, the factor VIII molecules separate out leaving other proteins and contaminants, including viruses, behind. Once again, the process appears to produce a safe product, especially when either solvent–detergent or heat treatment is used to mop up any residual virus.

As suggested in the final description, combinations of these methods can be used, resulting in a 'belt and braces' approach to concentrate safety. However, it will only be with the general adoption of entirely synthetic products that freedom from human viruses can be assured. Until this happens the transfusion services must keep *all* the safeguards in place, and doctors prescribing blood and blood products must tailor their recommendations to individual patients. The checks used to ensure the safety of patients are summarized in Table 11.1.

Table 11.1 Methods of ensuring the safety of human blood products. Volunteer, non-paid donors are preferred to paid donors for both single and multiple donor products. In the case of some blood products, like the immunoglobulins, the manufacturing process itself results in removal of virus. Viral inactivation methods using solvent–detergent technology is now available in some countries for single donor products. Most manufacturers now use at least two dissimilar methods to try to ensure that any viruses present in the source plasma are killed.

Blood product	Safety checks
Single donor products:	
Red cells	Exclusion of high-risk donors
Platelets	Hepatitis and HIV testing of each
Fresh frozen plasma	donation
Cryoprecipitate	Follow-up of recipients
Multiple donor products:	
Albumin	Exclusion of high-risk donors
Factor VIII concentrates	Hepatitis and HIV testing of each
Factor IX concentrates	donation
	Viral inactivation
	Follow-up of recipients

12

Side effects of treatment, 2: HIV infection

This section contains practical information about the human immuno-deficiency virus (HIV) and acquired immunodeficiency syndrome (AIDS). Within it are the answers to some of the very sensitive and private questions asked by people with HIV infection. I hope that anyone who feels uncomfortable with them, will seek the help of someone they can trust. One of the most wounding aspects of HIV infections can be loneliness, and the more people talk out their worries in confidence with others the better.

AIDS was first described in the United States of America in 1981. Although initially associated with male homosexuality in Western communities, AIDS is predominantly a heterosexual disease which is devastating Africa and other countries in the developing world.

In 1982 the first link with blood transfusion was found, and between 1983 and 1984 the cause of the disease was established as HIV. In 1984 a test for viral antibody was introduced, and the virus was shown to be heat sensitive provided that enough heat was applied to blood products for a sufficient length of time.

The extent of the infection in the haemophilic population had been realized by 1985. Since then a further virus has been found to be capable of transmission by blood products. The two viruses are called HIV 1 and HIV 2. To date very few people in the world are known to have HIV 2 infection, which seems to be less of a problem than an infection caused by the first virus identified. As our knowledge grows new varieties of the virus will no doubt be found. The important point is that this family of viruses are very susceptible to a variety of everyday factors capable of destroying them. These include soaps, detergents, and solvents as well as heat. When people have become infected more recently than 1985, for instance in France or more recently still in Germany, it has been because of human error and not because the stringent safeguards now in place to make clotting factor concentrates virally safe are at fault.

The various methods of viral inactivation are described in Chapter 6, and referred to again in Chapter 11. Thankfully, since adequate heating and other methods of killing viruses in blood products were introduced, *there have been no further cases of infection*. So, whilst everyone recognizes that no human blood product can be guaranteed to be completely safe, as far as AIDS is concerned the future looks very bright indeed for the new generation of people with haemophilia.

The virus

AIDS is caused by a virus which reduces the body's ability to resist invasion. The virus does this by attacking the white cells of the blood, and in particular by targeting one family of these cells, the lymphocytes. Within this family is a cell called a T-helper cell also called a T4 or CD4 cell (see background information). This cell is rather like the conductor of an orchestra; it is central to the many activities which provide the usual response to invasion of potentially harmful organisms. Without a conductor an orchestra falls apart. Without the T-helper cell so does the body's defence, or immune, system.

Background information
White cells and immunity

There are around 5000 white cells in a millilitre of blood. There are several different types, each with a specific task in helping to protect the body. In adults the commonest white cell, or leukocyte (*leuko* = white, *cyte* = cell) is the granulocyte, so-called because of the granules it contains. Granulocytes are like amoebae, the small single-celled organisms studied under the microscope by schoolchildren. As well as being carried round the body passively in the bloodstream, granulocytes are able to attach to and creep along surfaces by extending pseudopodia (false feet). Their ability to do this and to change their shape allows them to find their way through the walls of blood vessels and into the tissues. Here they scavenge foreign and waste materials. These refuse collectors of the body are stimulated especially when a bacterial attack is mounted. As bacteria invade they become coated with antibodies which are recognized by the granulocytes. Using a combination of chemical warfare and an ability to literally eat their opponents the granulocytes fight the infection, many dying in the process. Masses of dead granulocytes with their ingested enemies form pus.

The other main type of white cell, commonest in children, is called the lymphocyte. Recent work has revealed a host of new information about how lymphocytes work, information which is at the core of our understanding about how the body responds to substances foreign to it, including transplants and cancers (Fig. 12.1).

In addition to the circulatory system, which carries the blood around the body, there is another system which carries a fluid called lymph. Lymph, which comes through vessel walls from the blood, is collected from the tissues, filtered through structures called nodes, and emptied back into the bloodstream in the chest from a tube called the thoracic duct. Lymph nodes are known to most people as 'glands'. Although sited throughout the body, they may be felt with especial ease at the sides of the neck, in the armpits, in the groins, and at the top front of the thighs. If an infection, for instance a boil, occurs in the territory of a set of lymph nodes they become enlarged and may be sore. The enlargement is called lymphadenopathy.

The lymph nodes contain nests of cells programmed to combat invasion. There are two sorts of lymphocytes and they are called T and B cells. T cells are cells that have received their instructions on what to fight in an organ called the thymus, which lies just behind the top of the breastbone. The thymus is big in young children and because there is a lot to learn about how to spot an invader, there are many lymphocytes. B cells are lymphocytes that are programmed by the bone marrow, and other tissues. When they come into contact with antigens (recognition sites on bacterial invaders) B lymphocytes develop into another form called plasma cells, and plasma cells produce a very potent defensive weapon called immunoglobulin.

The immunoglobulins protect the body from mass bacterial invasion and toxins produced during bacterial warfare, and are responsible for preventing second infections with some viruses. The T lymphocytes have a different role. They are involved principally in the recognition of infections inside cells, for instance infections with viruses and some fungi and parasites. T cells protect either by destroying infected cells or by walling off infected areas with the help of other white cells. They are primed to recognize minor changes in the tissues, and it is the very effectiveness of this recognition that makes them responsible for the rejection of kidney and skin grafts from unrelated individuals.

The complex workings of T and B cells, and of the defence system of which they are a part, are termed collectively the immune system, and the science of their study is immunology (*immunis* is the Latin word for 'exempt' or 'secure').

It is obvious that everyday life produces multiple challenges to our immune system and that, sometimes perhaps whilst it learns about a previously unknown invader, it appears to fail us. It is when its failure is prolonged that real problems can arise. Such a situation may be produced on purpose with drugs in order to trick the body into accepting a transplant, or rejecting a cancer. However, it may also occur as the result of overwhelming invasion or continued bombardment by the invading forces, and there are also relatively rare inherited disorders of

Background (cont.)

immunodeficiency. In some of these cases both T and B cells are affected. More commonly a deficiency only affects one line, or part of one line, of the immune system. The acquired immunodeficiency syndrome, AIDS, is primarily an example of failure in T-cell function. Because B cells are less affected, bacterial infections in AIDS are relatively uncommon and patients are most likely to suffer from viruses, fungi like candida or 'thrush', and other invaders which are usually held at bay by the T-cell system.

The particular member of the T-cell family principally affected by HIV infection is the T_4 or T-helper cell. It is also referred to by a recognition site as a CD_4 (cluster determinant 4) cell.

The HIV virus is shown in Fig. 12.2.

In order to gain access to its target cells it must first enter the body. This entry is achieved by the injection of infected material, either by sexual intercourse, or by contaminated needles or other equipment, or by contaminated blood transfusion. In addition, a woman who is infected can pass the infection to her unborn child.

It follows that no one who is infected with HIV, or their sexual partners, should give blood or donate organs. They should not carry organ donation cards of the sort that alert doctors to possible help with a human transplant following an accidental death.

Once in the body the virus locks itself into recognition (antigen) sites on the surface of the target, T4 cell. The docking sequence complete, the virus is then able to discharge its cargo directly into the cell.

The viral cargo is RNA. Normally, RNA carries instructions *from*

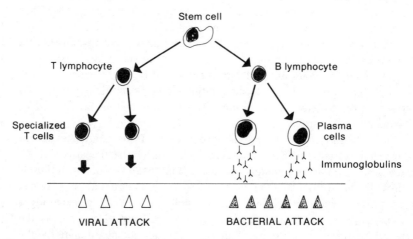

Fig. 12.1. Cell wars: the body's lymphocytes develop into two battle fleets to repel invaders.

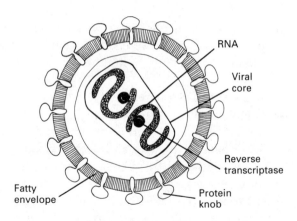

RNA

Viral
core

Reverse
transcriptase

Fatty
envelope

Protein
knob

Fig. 12.2. The human immunodeficiency virus (HIV) which causes AIDS. Once attached to a white cell membrane the virus hijacks the cell using its two strands of RNA and the enzyme, reverse transcriptase. The fatty envelope of the virus is easily disrupted by soap, detergents, disinfectants, solvents, and heat; as a result HIV infection is not contagious.

the genetic blueprint DNA into the cell factory. HIV reverses this sequence by injecting a viral RNA messenger, complete with all the instructions for making more virus, together with an enzyme called reverse transcriptase. This enzyme hoodwinks the cell factory into making DNA from the RNA intruder. The new DNA then slides neatly into the blueprint stored within the cell's nucleus. Once in the person's genome the viral blueprint can be activated at any time to tell the cell factory to make more virus. Because this highjack of normal cell mechanisms works backwards—RNA to DNA instead of the other way round—HIV is called a retrovirus.

As the invasion proceeds more of the target cells die and remaining immunity gradually disappears. The triggers for the release of new virus from the cells of the host include age—old people are more likely to have symptoms of disease earlier than young people—and the presence of other conditions known to lower immunity.

Thankfully the virus has an envelope, or membrane, which contains fat. This means that it is a very fragile beast which is easily destroyed by soap or detergents and by chemicals including bleach, disinfectants, and solvents. It is also susceptible to heat and certain forms of light. When clothing is contaminated the virus is, for instance, destroyed in an ordinary household hot wash.

Its fragility means that it cannot be passed from person to person in a casual way by everyday contact. The virus is *not* spread in the ordinary activities of home, school, employment, sports, pubs, public transport, or places of entertainment. It is *not* spread by using the same toilet as somebody who is infected. It is not spread by kissing or by sharing the chalice during Communion in church. In other words HIV *infection is not a contagious disease.*

No one has developed AIDS as a result of any contact other than the direct injection of infected material into the body. Naturally this is most likely to occur during sexual intercourse or pregnancy, or during the sharing of needles or equipment with an infected person. *Very rarely* needlestab or other injuries with infected sharp instruments have resulted in health care workers becoming HIV positive. In addition, a few people with open wounds—for instance weeping eczema of the hands—who have not taken normal hygienic precautions when handling infected material have become infected. A handful of cases in families may have resulted from the sharing of razors or toothbrushes contaminated with blood.

At the time of writing world-wide surveillance had only revealed 140 cases of transmission or possible transmission to health care workers from infected patients since the epidemic began. In contrast, there is no direct evidence of transmission from health care workers to patients. Despite intensive research and widespread publicity the case of a Florida dentist with AIDS and his infected patients remains a mystery.

When I examine anybody with HIV infection or AIDS I take no special precautions other than washing my hands carefully with soap and water before and after the examination. I only wear gloves in the course of procedures which would normally require them, or when taking blood or giving an injection. If I have a fresh cut I simply cover it with a waterproof dressing.

These precautions are increased in operating theatres, delivery suites, and dental surgeries because of the danger of infected blood getting into the eyes, nose, or mouth of staff. So do not be alarmed when doctors and nurses wear goggles in addition to the usual operating theatre clothing of masks, aprons, gowns, boots, and gloves in these situations.

If blood or other fluids are spilt in the consulting room or the ward, the mess is cleaned up using the disinfectant hypochlorite which is, for instance, present in the ordinary household bleach, Domestos. A 1 in 10 dilution of Domestos contains enough of the active ingredient (chlorine) to deal with spills or gross contamination. Surfaces and soiled articles can be cleaned satisfactorily with a 1 in 100 dilution of Domestos. It is sensible to wear disposable gloves whilst cleaning up.

The usual way that needlestab happens is during the resheathing of a used needle. Because of this needles should *never* be reintroduced into their protective sheaths. All needles, including small-vein sets, should be discarded into a needle disposal box immediately after use. This box should never be allowed to fill to more than two-thirds of its capacity because injuries have occurred when used equipment has been forced down into an overfilled container. On no account should any sharp equipment, including needles, be disposed of in an ordinary household refuse collection. This may expose refuse workers, or even inquisitive children, to potential contamination.

If anyone other than someone with haemophilia treating himself accidentally punctures their skin with a used needle they should:

- encourage bleeding from the puncture site; then
- wash the site liberally with soap and water; then
- swab the site with antiseptic and cover it with a waterproof dressing; and
- report the incident to their doctor within 24 hours.

The chances of becoming infected are very remote. There is a school of thought that a short course of zidovudine may prevent infection becoming established if substantial accidental exposure, for instance the actual injection of infected blood, has occurred. To be effective such a course should probably be started within 1 hour of exposure. In any case people who have been exposed may wish to have an HIV antibody test performed, with a repeat some 3–6 months later in order to indicate absence of infection. In order not to be put into the position of admitting to an HIV/AIDS test by the insurance companies, I suggest that a blood specimen is deep frozen for future testing if necessary. In this way the person concerned can truthfully say they have not been tested for the disease when applying for insurance.

The course of HIV infection

Improvements in the prevention and treatment of opportunistic infections and in the maintenance of good health have already influenced the natural history of HIV infection. This better management has gradually enhanced both the length and the quality of life until it is now impossible to predict the outcome in all infected people. Indeed, present evidence suggests that there may well be people with HIV infection who will never develop AIDS. It is very important to remember this because it means there are grounds for real hope, and therefore encouragement to do everything possible to keep people well and cheerful.

The natural history of HIV infection is shown in Fig. 12.3.

A few weeks after infection a flu-like illness, which has been likened to glandular fever, may occur and within three months most infected people will have become HIV antibody positive. This is called seroconversion. The time interval between infection and antibody positivity has been called the 'window', and there is a danger that someone who has been infected (but who does not realize it) can give blood in this interval, before the positive antibody test which would lead to the automatic rejection of his or her donation. At the current very low rate of HIV infection in the UK this risk is very remote.

The time interval between infection and the development of the illness we call AIDS is extremely variable.

For this reason doctors no longer think in terms of the two distinct phases, HIV infection and AIDS. Instead they regard the whole disease as a spectrum or continuum throughout which both the quality and quantity of life can be altered in the patient's favour by appropriate intervention. This change in attitude is important because, from the patient's viewpoint, a diagnosis of 'AIDS' is no longer the critical watershed between success and failure, life or death. The spectrum attitude also allows for the fact that there is no such thing as an 'AIDS test'. The diagnosis of AIDS is only really of importance for the purposes of following the epidemic, and not for the well-being of the individual. It is made on one of two definitions, depending on the authority responsible for AIDS surveillance. For instance, at present the definition of AIDS in the United States of America includes a fall in T4 cells, whilst in Europe the diagnosis is a clinical one triggered by the development of one of a long list of diseases linked to HIV infection.

As time passes from the original description of AIDS in 1981, we learn more about the course of the infection, and the average time lengthens. For instance, when I wrote a booklet for the UK Haemophilia Society in February 1985, the average interval was thought to be 28 months, with

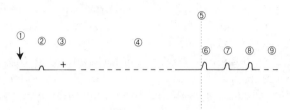

Fig. 12.3. The natural history of HIV infection. Infection (1) is followed in some people by a flu-like illness (2). At (3) the HIV antibody test becomes positive. There then follows a long, quiet period when the infected person feels and looks healthy (4). (5) AIDS is diagnosed when the first illness (6) to signal immunodeficiency occurs. Other illnesses may follow, (7) and (8). Between these illnesses, people can expect a good quality of life which can be extended with zidovudine and other medicines (9).

a range of between nine months and six years, and it is now twelve years. During the interval most people feel and look perfectly well, and even a detailed physical examination will reveal no evidence of disease. Therein lies one of the major dangers of HIV infection, because once infected a person is capable of infecting others. Hence the World Health Organization campaign, echoed in many countries, to educate the public and to try and prevent the spread of the disease by reducing the number of sexual partners, encouraging the use of condoms, and curtailing the sharing of needles in intravenous drug abuse.

Early in the epidemic doctors recognized a phase in HIV infection they called ARC, the letters standing for AIDS-Related Complex. With the recognition that symptoms and signs of infection can occur sporadically, and do not necessarily herald the imminent development of AIDS, the term is no longer appropriate. The symptoms and signs are usually non-specific, and of the sort that everyone is used to when they are off-colour for a while. They include fatigue, weight loss, muscle weakness and wasting, an abnormally dry skin often with severe dandruff, night sweats, intermittent fever, headaches, cough, diarrhoea, and swollen lymph nodes ('glands') in the neck, armpits, and groin. In HIV infection these signals of ill health tend to recur or persist, and in some cases can be profound enough to interfere with work.

One event which seems to be peculiar to haemophilia is infection in a major joint, called septic arthritis. Initially the diagnosis of this can be difficult because the presentation is the same as for a haemarthrosis—the joint is swollen and painful, and may feel hot. One clue is that bleeding into the joint should respond to blood product therapy. If it does not, infection should be suspected. Sometimes the only way to be sure is for the doctor to put a needle into the joint space. If pus is present the bug responsible can be cultured and the appropriate antibiotics given. Septic arthritis is more likely to occur in an artificial, replacement joint than in a joint which has not been operated on in the past.

Obviously, if any of the symptoms are particularly severe or troublesome the advice of a doctor should be sought, and it is wise to do this anyway if any of them persist for more than a few days. Especially important is to have a check-up if shortness of breath occurs, with or without a cough. This is because it may be a sign of *Pneumocystis carinii* pneumonia which is common with HIV infection.

The occurrence of one of more of these symptoms can be very debilitating, and it is wise to do everything possible to maintain a good bodyweight, and not let everyday activities and social contacts fall away. All the symptoms occur in conditions other than HIV infection—for instance, fatigue is a feature of anaemia due to blood loss—so do not automatically assume that they are inevitably linked to it. However, three other conditions are commonly linked with HIV infection. They are thrush, recurrent cold sores, and shingles.

Thrush

This is the common name for an infection caused by a fungus called *Candida albicans*. Albicans means white, and thrush is easily recognized by the white plaques it produces on the membranes of the mouth, gullet, or genitalia. Because the plaques stick to the tissues they are irritant, and a heavy growth can be very sore. In the mouth or gullet thrush can interfere with eating by making swallowing painful. Thankfully the infection is usually very easy to deal with, especially if treated early. A number of medicines may be prescribed, but the best are nystatin for first infections, and fluconazole or ketoconazole for more resistant growth. Anyone with HIV infection who has had an episode of thrush should keep an antifungal medicine at home so that a second episode can be dealt with promptly. Sexual partners may also need treatment to reduce the chances of reinfection.

Cold sores

Similarly, the specific medicine for cold sores should be readily available. These sores are caused by the virus herpes simplex, which is kicked into action by lowered immunity or, in susceptible people, by sunlight and, of course, colds. The appearance of a cold sore is preceded by a tingling feeling in the skin, and if acyclovir (Zovirax) ointment is rubbed in at this stage the full lesion will not develop. If resistance to local treatment develops acyclovir tablets, taken at the first hint of trouble, will abort the episode.

Shingles

Acyclovir is also the drug of choice in shingles. Shingles, or herpes zoster, is caused by the chickenpox virus. Usually resulting in a band of small blister-like spots (like early cold sores) on one side of the trunk, it can occur on the face. Here it is particularly important to seek early advice, because the infection may involve the eye. As with cold sores, shingles is preceded by a tingling feeling and reddening of the skin. At this stage high-dose acyclovir by mouth can halt, or at least slow, progression. If the lesions have already developed by the time treatment is started intravenous medication may be necessary in hospital.

Follow-up

Because so much can be done to treat most opportunistic infections, especially if the treatment is started early enough, regular follow-up is essential. Table 12.1 shows one checklist used when seeing people with HIV infection. All the checks are straightforward and the investigations routine. As far as possible, people are seen with their wives, friends, or families, but a chance is always given for them to be seen alone for a while as well. This means that there are always opportunities for personal questions to be answered in the light of up-to-date knowledge. Advances in our understanding of HIV infection result in a continued upgrading of advice to patients and families, and this advice is best passed on in the privacy of a consulting room.

Knowledge of infection and the development of symptoms, especially weight loss and muscle weakness and wasting despite attempts to eat well, can be profoundly depressing. In these circumstances it is sometimes worthwhile to try an anabolic steroid like stanozolol (Stromba), or the hormone preparation megestrol acetate (Megace). Anabolic steroids are protein builders, hence their illicit use by some misguided athletes. Megace, which is usually prescribed for the treatment of cancers of the breast or womb lining, has weight gain as a side-effect and this can be useful in HIV infection.

Like zidovudine, the start of which is also usually associated with weight gain, Stromba and Megace can produce a feeling of well-being and self-esteem, and boost body image. Obviously, they should only be taken under careful medical supervision.

Table 12.2 provides suggestions for medicines that should be readily

Table 12.1 Routine follow-up checklist

History	Checks on appetite, fatigue, shortness of breath, skin rashes or dry skin, dandruff, night sweats, diarrhoea, headache, thrush in the mouth, cold sores.
Physical examination	Checks on height, weight, and on the normality or otherwise of the eyes, mouth, skin, lymph nodes (glands), chest, and abdomen.
Tests	Checks on the urine, and on the blood count, immune status, and liver function.

Other tests, for instance X-rays, are dictated by the findings. Follow-up should be at least every three months for people without symptoms.

As well as the individual history of physical health, other questions relating to general well-being and comfort, and sexual and family health may be explored.

Table 12.2

Medicines	Indications
Trimethoprim trimethoprim–sulfamethoxazole (TMP–SMX), dapsone, dapsone pyrimethamine, or pentamidine	Prophylaxis to prevent *Pneumocystis carinii* pneumonia
Nystatin or fluconazole	Thrush
Acyclovir cream or tablets	Cold sores or shingles
Miconazole/steroid cream	Dermatitis
Betamethasone scalp lotion	Severe dandruff

It is also useful to have remedies for dry skin and soreness of the mouth or throat handy. We use arachis oil and providone cream (Oilatum), and a benzydamine mouth wash (Difflam).

Some haemophilia centres arrange regular intravenous injections of gammaglobulin for patients, particularly children.

The choice of *Pneumocystis* prophylaxis is one for the individual and his doctor.

available to people with HIV infection so that they can start treatment themselves as soon as they recognize the early symptoms of disease.

Follow-up is, of course, imperative for anybody on one of the antiviral medicines such as zidovudine, in order to spot and deal with any side-effects early. Because these medicines must be evaluated very carefully to ensure that they are both effective and safe, many patients are generous enough to join clinical trials, sometimes involving a placebo (an inert substance). I believe that these trials are essential because they are the only way in which we can judge whether or not a medicine is effective.

Testing partners and children

Now that so much can be done to help HIV antibody positive people live a good quality of life for prolonged periods, there is a sound argument for offering the antibody test to their sexual partners, and to babies born since the beginning of the epidemic. People offered the test should obviously think very carefully about it, because confirmation of positive HIV antibody status means that they are infected and will have to face all the hardships associated with this diagnosis. They will be aware that they can infect others, and that their options for marriage and parenthood may be lost. They will not be able to obtain life insurance cover and will, therefore, have a limited choice with regard to mortgage arrangements. They may be restricted in their plans to travel or work in other countries.

On the other hand, if the test result is negative the knowledge brings great reassurance.

Whatever the result it ends uncertainty and, with time, allows people to gain control over their lives again. My advice is that it is usually worthwhile knowing, and this is especially true in the case of children, and casual sexual partners who may develop other relationships or become pregnant in the future.

When somebody has been exposed to the possibility of infection they should be offered HIV antibody testing straightaway, and again three to six months after exposure. Their confidentiality should be strictly respected, and nobody else told of the reasons for requesting testing or of the result without their permission. As discussed on p. 239 some people, fearing that they may have been exposed to the risk of infection, prefer a specimen of blood to be taken and deep frozen for later testing if the necessity arises.

Telling other people

Knowledge of HIV infection and the public attitude towards it can very easily lead to isolation. People are frightened of invasions of privacy and of ill-informed and sometimes cruel comment. They may be bitter and angry and they may deny that there is anything wrong with them. This denial can be valuable initially because it affords them protection and helps stop anxiety or depression which might otherwise threaten to overwhelm them.

At some stage, however, everybody needs to talk to somebody they can trust. This can be done with a member of staff in the centre, or with the family doctor, or perhaps with somebody from the voluntary societies. However, what is really needed is the informality and comfort of everyday conversation with family and friends. My advice to people is that they should talk as much as possible and try to bring themselves to telling close friends the truth, unless there are real grounds for not doing so. I think that this is even more important in the case of families, because the hiding of the diagnosis within a close-knit family must be attended by stress, and because it means that those with the knowledge have to be constantly on their guard. It can also be very hurtful to those who have been kept in the dark. It takes courage to share the diagnosis but my experience is that the sharing is almost invariably worthwhile.

When children ask questions it is wise to answer them truthfully in a way that is in keeping with their age. Most parents are eventually able to bring themselves to tell a child that he or she is infected in the privacy of the home, but it is occasionally easier to ask a doctor to do this. When this has happened to me I have always found that the boy or girl has known about the diagnosis for some time anyway. One youngster said that he had not wanted to frighten his parents by telling them! It is hard, but once the truth has been shared, both child and family can begin to relax and get on with living.

Opportunistic infections

When someone's immunity fails, invaders take advantage and gain access to the body. In people with haemophilia and HIV infection the commonest of these are thrush, and a tiny organism called *pneumocystis carinii*. As we have seen thrush can be particularly troublesome in the mouth, and pneumocystis can cause shortness of breath.

All these conditions are treatable and, once they are under control, people with them can expect to continue to live enjoyable and fruitful lives. It is easier to gain this control if treatment is started early, and in some cases prophylaxis is worthwhile to prevent illness in the first place. This is especially so with pneumocystis where there is a choice of prophylactic medicines including trimethoprim, dapsone, and inhaled pentamidine, all of which can be taken regularly at home (see Table 12.2).

The invasion of white cells by HIV allows the virus to find its way to all parts of the body, including the brain. Here it can cause changes which, in their extreme form, may result in dementia. This condition is very like that seen in elderly people, one of the most noticeable features being forgetfulness. Thankfully, earlier predictions that many of those infected with HIV would develop profound dementia have not been realized. In addition to HIV, some opportunistic infections can affect the brain function. Like the conditions mentioned above, many of these are treatable, especially when diagnosed early.

The treatment of HIV infection and AIDS

At the time of writing many drugs are under development or in clinical trial. The first drug to be used extensively was zidovudine (AZT). There is still debate about whether zidovudine is more effective when given early in the course of HIV infection before symptoms have developed, or not. A combined British and French study (Concorde) has suggested that there is no benefit from early treatment, whilst other trials have given contradictory results. My current view is that zidovudine should probably be reserved until symptoms appear. There is then sufficient evidence of a beneficial effect in many infected people, and its use can help give added months or years of good quality life without the worry of inducing early resistance to its effectiveness.

Many other drugs are currently in clinical trial and some look very hopeful; all require that people on them are followed up regularly. Some may work better with other drugs, in 'combination chemotherapy'. Many cancers, including leukaemia, respond better to multiple treatment of this sort, which targets different parts of the life cycle of cells.

So the treatment of HIV infection and AIDS is presently one of maintaining a good quality of life by using zidovudine until an alternative, and less toxic medicine, is proven, and then making the switch. I explain this to my patients by saying that they are like frogs on a lily pad, and that as new lily pads become available they can hop from one to another!

None of the drugs under investigation, including zidovudine, are likely to cure AIDS; they simply allow the life of someone who is infected to continue at a level as near to normal health as possible. There is nothing to be frightened of in this concept because it is commonplace in medicine. After all we cannot yet cure haemophilia (except by the radical surgery of liver transplant), but we can treat it very effectively.

It would be good to suggest that a vaccine against HIV was on the horizon, because that would of course bring great cheer to couples aware of the infection in only one partner. Unfortunately, because there are so many differences in the viral envelope, an early vaccine which can recognize most, if not all the varieties, and protect against them looks remote. But strange things happen in medicine and no one should ever give up hope. Serendipity, the creation of happy and unexpected discoveries by accident, has been known to confound the scientists before!

Dying with HIV infection

People with severe haemophilia have always faced the possibility that an accident or particularly severe bleed may mean that they die earlier than usual. They have also weighed the consequences of modern therapy and, with their doctors, have participated in the striking of a balance between the probable benefits and the possible costs of their treatment. The appearance of AIDS has tipped that balance so that the lives of many more people are ending prematurely. Their deaths can be clouded by misinformation and myth, and it is the purpose of this section to reassure those infected with HIV, and their families and friends.

Firstly, there is nothing unusual in the death of someone with AIDS; their dying is as dignified and as peaceful as any non-accidental death. People do not die 'from AIDS', they die from one of the diseases associated with the lowered immunity of HIV infection. None of these diseases are new, all were seen and cared for by doctors and nurses before AIDS appeared. So there is no need for people, whether staff or families, to be burdened with additional fears at this difficult time. There is no reason why someone should not be nursed at home rather than in hospital during their last illness, but this is a matter for the patient, his family, and his family doctor in consultation with the centre staff.

Many people with haemophilia feel safer in hospital. Often it is better to be in hospital where 24-hour nursing care is provided and any injections of sedatives or pain-killing drugs can be given when they are needed. Whatever decision is taken it is of crucial importance that arrangements are made for relatives and friends to stay and care for their loved one, without palaver and at any time. Easy access to a telephone, and to radio, TV, and video should be provided. Favourite foods and drinks, including alcoholic drinks, will often be welcomed.

When I talk to people with HIV infection one of their great fears is that they must die alone in the sterile confines of a special room in hospital. They need reassurance that this is not so. People with AIDS can be cared for in ordinary hospital wards, and although some will prefer the privacy of a room or cubicle to themselves, others will be cheered by the opportunities of having other patients to talk to when visitors are not with them. Most of the diseases associated with HIV infection do not require barrier nursing, so there is no need for visitors to wear special clothing. Gowns, masks, gloves, overshoe covers, and other forms of protective dress may be needed in an operating theatre, but have no place in the nursing of terminally ill people with AIDS, unless a particularly infectious disease is complicating matters. Nor is there any need whatsoever to caution ordinary physical contact; families should be at ease to hold and to hug and to touch and to kiss. And, of course, children and grandchildren must be allowed to visit whenever the family feel it appropriate.

Another unnecessary fear is that strange rituals follow the death of someone with HIV infection. Some people have said they have been frightened because they thought that no one was allowed to touch or kiss someone they loved who had died with AIDS. They thought that it was obligatory to have a cremation, and perhaps worst of all, that it was obligatory to have AIDS or HIV disease written on the death certificate. What are the facts?

Before a body is released to the undertakers, relatives and friends must be given time and privacy to take their leave. This is very important because, for reasons explained below, the body sometimes cannot be viewed again. Within some parts of the United Kingdom, because it has been decided that HIV infection and hepatitis B infection should be managed in the same way, at a suitable time after death the body is placed carefully into a body bag which is then sealed. The reason for this is that those handling the body could have open wounds or sores and could inadvertently be exposed to infection. So, although the chances of this happening are very remote, some authorities think it sensible to offer them this simple and straightforward protection. What it means to the family is that the body cannot lie openly in a Chapel of Rest or at home. People who die with AIDS may be buried or cremated, that is their and their families choice.

In the UK it is not necessary for death certificates to reveal AIDS or HIV disease. The details needed for disease surveillance may be recorded elsewhere, and all the certificate need reveal is the immediate cause of death which may, for instance, have been pneumonia. For reasons of confidentiality we try to avoid open inquests (unless the family request this form of investigation, or there are other legal requirements), but the Coroner is always informed privately of the death with AIDS of someone with haemophilia.

HIV infection and sexual health

One of the saddest aspects of HIV infection is the fact that it is transmitted via sexual intercourse. This means that an infected partner, whether male or female, can transmit the virus to a partner who is uninfected.

Within the body the concentrations of the virus are strongest in the semen and vaginal secretions, as well as in the blood. Two factors are thought to help sexual transmission—high levels of virus and breaks in the skin of the genitals. High levels of virus can be expected shortly after infection and towards the end of the spectrum of infection. At these times people are said to be viraemic.

It follows that, between these times, infected people are less viraemic and possibly less likely to transmit HIV during sexual intercourse. This may be the explanation why some couples, in which one partner has been an infected man with haemophilia, have had normal, uninfected children. Thankfully, the wife has, in most of these families, also remained uninfected. However, it must be stressed that to have unprotected intercourse with anyone who is infected, no matter what the stage of infection, is like playing Russian roulette. Doctors working with people infected with HIV are all aware of cases of infection occurring as the result of single acts of unprotected vaginal intercourse. If this happens at the same time as conception our current knowledge is of a 15 per cent chance that the baby will be infected too. This risk is thought to be much higher in developing countries.

However, despite being fully aware of these risks a few couples do continue to make a conscious and very human decision to have children. Couples making this decision should try to pluck up the courage to discuss it with their doctor who will be able to give them up-to-date, factual information and to arrange for expert care during pregnancy and delivery. My advice to them is to abstain from sexual intercourse or to use condoms other than at the time of ovulation in order to reduce the risk of viral transmission. The time of ovulation can be estimated by using one of the kits available from pharmacies, or by applying the principles described in the literature on natural family planning. Once a woman knows she is pregnant then abstinence is probably wise so that neither she nor the child can be infected before delivery.

Other than blood, semen, and vaginal secretions the only other fluid known to have transmitted HIV is breast milk. We know this from a few cases in which a mother has become infected after delivery as a result of a contaminated blood transfusion and has then infected her baby whilst breast feeding. For this reason we discourage mothers thought to be infected from breast feeding their babies. Probably no harm would come if an infected mother wanted to feed an infected baby, but it is difficult to test infants until they are around 15 months old. This is because HIV antibody from the mother crosses into the baby during pregnancy, and tests on the child simply reflect those in the mother until her antibody has had time to clear from the baby's bloodstream.

Some couples have asked about the possibility of separating seminal fluid that contains white cells which might be infected from sperm, and then using artificial insemination to achieve pregnancy without risk of infection. Unfortunately, HIV seems to be capable of invading the sperm themselves and so this way of guaranteeing an uninfected wife and baby is not possible.

The only way we know to prevent sexual transmission of HIV, other than abstinence, is to ensure that a barrier separates the tissues of the partners. Condoms and their use are described in Chapter 15. Provided that they are used correctly, they will provide extremely good protection during vaginal intercourse. They are not recommended during anal intercourse, which is anyway a very unsafe practice if one partner is infected, because the delicate lining of the lower bowel is readily injured leaving the way open for the virus to gain easy access to the bloodstream.

Protection is increased if the spermicide nonoxinol 9 is used in addition to the condom. Nonoxinol 9 is available in a variety of preparations including pessaries, foams, creams, and jellies. Remember that none of these are effective on their own, and that a condom must be worn with them.

In addition to these precautions women who are already on the pill are advised to continue to take it. This is because, *although the contraceptive pill affords no protection against HIV,* it does provide the best protection against pregnancy. Couples who already have children may wish to consider sterilization.

Despite the risks and the obvious pressure to counsel abstinence I am very reluctant to do this when talking to couples with stable relationships. Many of them will have enjoyed a fruitful and happy sexual partnership for years, and the emotional break from such a relationship can be very damaging and hard to bear, especially at a time so beset with isolation and loneliness. Using condoms at such a time can be very depressing,and some couples have said it is a constant reminder of what has happened to them. But it does provide for the protection of a loved one and therefore has to be recommended. At the present state of our knowledge this is the only way we know to provide this protection. Hopefully, the poverty of this option will be eased in the future. Unfortunately, this poverty extends to the words we use as well, because they are either very clinical or very coarse. Someone should invent a new language of love! Until they do the next paragraph has to be clinical—except for the last sentence . . .

With the possible exception of oral sex other forms of sexual activity are safe, healthy, and fun. Oral sex has to be a possible route of transmission especially if ejaculation occurs in the mouth. Similarly, caution is suggested with mouth to vulval or vaginal contact, and barrier protection has to be recommended. Mutual masturbation, hugging, kissing, massage, and just simply holding one another are all fine—and what about baths, showers, or saunas together? And despite all the gloom, don't forget courtship—flowers, music, wine, chocolates, sharing each other, and days in the sunshine and fresh air can still lift the spirit!

Advice to youngsters

Young, sexually aware people who know they are infected with HIV have an unenviable responsibility. On the one hand they want to fall in love and grow to sexual maturity and perhaps marriage and parenthood just like anyone else. On the other hand, they know that sexual intercourse will expose their partners to the risk of viral transmission. To try and help them find a way through the dilemma of living with HIV as they grow up, the advice we give them is straightforward and unequivocal.

First, they need to be told that they are human and that their sexual feelings are normal. No one can expect them to opt out of life and become hermits.

Like all humans they have a responsibility to themselves and to others. In the case of shared sexual activity this responsibility must be to protect the other person from infection.

The only sure way of doing this has to be the avoidance of sexual intercourse.

It would, of course, be naive to imagine that all relationships are going to remain chaste because of HIV. So youngsters must be armed, not just with the theory of protected intercourse but with the means of practising it. This means an easy, unthreatening, and unembarrassing access to condoms, and to preparations containing nonoxynol 9, which are available without prescription in chemists.

Unfortunately the use of condoms is not considered to be very macho in some parts of our society. So our advice is to make use of all the publicity about AIDS in the population, and to stress that condoms are there to protect *both* partners—not just against AIDS but against other sexually transmitted diseases as well. The fact that the girl is on the pill does not weaken this argument. If she is not on the pill, it strengthens the argument because the condom is then a contraceptive too.

So, what happens if a relationship blossoms. Well, there has to be a time to tell, both about haemophilia and about HIV, and there is no easy way to do that. It helps if the couple know that they have quick and confidential access to someone who is both sympathetic and knowledgeable about HIV. Most haemophilia centre staff, and many family doctors have the knowledge, as do an increasing number of teachers and religious advisers. And there are many volunteer counselling organizations including, in the UK, the Terrence Higgins Trust as well as sexually transmitted disease (STD) or genito-urinary medicine (GUM) clinics in all major towns. It's a question of summing up courage and being willing to talk. We find these consultations easy, unembarrassing and, more often than not, lightened by a gentle humour.

HIV infection and education

Young people with HIV infection must be encouraged to go to school or college and join in all activities including sports, clubs, discos, and outings in the usual way. They should go all out to compete with others in competitive exams and career planning.

In the UK there is no requirement for teachers to be told about the infection, because normal hygiene and first aid are sufficient to cope with HIV, and there is no danger to other children. Sometimes teachers ask me to confirm that a particular child is HIV negative but I refuse on the grounds that to identify the uninfected would, by exclusion, expose those children who are infected.

Occasionally parents decide that a teacher should know the diagnosis and, of course, this makes sense if the child needs special help or counselling. Haemophilia centre staff can help here, by giving the teacher factual information, reassurance, and support. Confidentiality and the child's right to privacy should always be stressed.

Infected boys going abroad on school trips should carry brief details of their infection, as well as the usual information about their haemophilia. This is a sensible precaution in case they become ill whilst away, and should ensure that they receive prompt treatment.

Recommendations for vaccination are given in Chapter 14. The only routine preparation which should be avoided in an infected person is BCG, the anti-TB vaccine.

Sometimes young people ask about risks associated with tattooing and ear piercing. These practices are safe provided they are carried out in a professional establishment, where strict precautions to prevent cross-infection with hepatitis and HIV are taken. Remember that professional tattooing is associated with bleeding and treatment for haemophilia may be needed!

Another practice, sometimes popular with children, is definitely not safe. In blood brotherhood or sisterhood blood is exchanged between children who have cut or scratched themselves as a mark of friendship. This ritual is highly dangerous because both HIV and hepatitis B may be transmitted.

HIV infection and employment

Some 16 per cent of men with haemophilia have always thought it better not to share their diagnosis with employers because of fear that they will lose their jobs. The same is true of HIV, and sometimes the man has to shoulder the burden of hiding both conditions from his workmates. In general, this practice is not recommended because of the enormous stress that may be involved, but also because of the genuine help and support which is often forthcoming from the most unexpected quarters when the diagnosis is shared.

However, the decision is very understandable in some jobs. For instance, a family who rely for their livelihood on the catering trade may fear that their clientele will disappear to eat elsewhere if they know that the father and head chef is HIV positive. Whether or not people at work know the diagnosis, two things are very important. Firstly, keep working if you can—especially if you enjoy it. Secondly, try to keep appointments for follow-up even if this means you have to miss the occasional half-day. It is very easy to deny the fact of HIV infection, and going to work can provide a very good excuse for not turning up at those essential follow-up clinics.

One further aspect of employment, and indeed of holidays, that needs mention is the question of travel restrictions. Several governments have introduced restrictions on people wishing to work in their countries, whilst others have made life difficult for infected tourists. Nowadays it is wise for families planning to travel or to work abroad to seek up-to-date advice from their centre or haemophilia society.

There is generally no need for anyone to disclose HIV infection to their employers. Nor is there generally any need to undergo testing for HIV in the course of a medical examination. HIV should not be a bar to employment, and does not constitute an adequate reason for dismissal in the UK. Responsible employers should be aware of the absence of spread by everyday contact, and of the need for good hygiene at work. First aid personnel will know about precautions for hepatitis B; precautions for HIV are exactly the same. Anyone coping with accidents should cover exposed cuts and abrasions with waterproof dressings and should wash their hands with soap and water before and after attending to someone with an injury. Mouth to mouth resuscitation has never been shown to transmit HIV and no one should have qualms about using it in an emergency.

A last word

In the past decade millions of words have been written about AIDS, and there are innumerable sources of reference and help. But all the paper does not fulfil the most important need of someone who is infected. This is for human contact, and opportunities to talk, and to laugh, and to cry in the peace of trusted relationships with family and friends. With this support, and the best medical care available, take every opportunity to get on with living. Keep active. Arm yourself with enough sleep, a good diet, and exercise. Keep working, but get away on holiday when you can. Take charge!

I have always encouraged people with haemophilia to put living first, with haemophilia firmly in second place where it belongs. The same is true of HIV infection and AIDS. You should be in control, not the disease.

13

Coping without treatment: the developing world

Haemophilia care in developing countries

Whenever I visit haemophilic families in the developing world I am both heartened by their spirit, and dismayed by the way many of them are subjected to misleading information about treatment. It cannot be over-emphasized that clotting factor concentrates are not the primary requirement for effective haemophilia management. Concentrates are always expensive, and often difficult to obtain. They divert both finance and attention from the basic care of many families to the interests of the few who are able to arrange access to them. So whilst it is true that home therapy and major surgical programmes in wealthier countries are now reliant on supplies of potent factor VIII and IX freeze-dried concentrates, it is also true that a good quality of life was enjoyed by many haemophilic families before their widespread introduction in the 1970s. Even today only around 20 per cent of the world's need for factor VIII is met despite the advent of recombinant products. So what can the majority of people with haemophilia do to help themselves? This chapter is concerned with some ideas for haemophilia health in the developing world.

Educate people about haemophilia

My colleague Dr Mammen Chandy, who works in India, believes that 50 per cent of the problems he sees in people with haemophilia could have been avoided simply by education. The comparative rarity of haemophilia means that most people, including doctors, will not be familiar with the dos and the don'ts of haemophilia care. It is therefore the responsibility of everyone concerned with haemophilia, and most especially affected families, to learn all they can about the disorder and to teach others the rules.

Here are some of the things that should **NOT** happen to someone with haemophilia:

- he should not be given aspirin, or any compounds containing aspirin;
- he should not be given tablets that contain non-steroidal anti-inflammatory agents for his arthritis;
- he should not be given medicines by intramuscular injection;
- blood should not be taken from the jugular or femoral veins for diagnosis, or for testing at follow-up;
- a child with haemophilia should not be removed from his parents when a blood sample is needed or treatment is to be given. Most procedures in the young child can be performed easily with him sitting on his mother's lap. Enforced restraint on an examination couch away from parents will provoke fear and bleeding, and discourage the reporting of accidents or bleeds in the future;
- joint aspiration should not be performed. If aspiration is thought to be essential in order to diagnose infection in a joint, it must only be carried out under full aseptic theatre conditions and be followed by clotting factor replacement cover;
- his joints must never be forced beyond the range of movement he is capable of when he is not bleeding. For instance, trying to force additional extension in an elbow or knee the range of which is already limited by haemophilic arthritis can cause both bleeding and further structural damage;
- he should not be subjected to dental extractions and planned surgery unless an inhibitor has been excluded by laboratory testing. Enough clotting factor cover should be available to control bleeding. Surgery performed with inadequate cover is more expensive both in terms of money, and in terms of patient health, than surgery performed with carefully planned cover. The use of simple blood products, together with antifibrinolytic drugs and immobilization of the wound until healing is advanced, reduce the need for scarce resource. Continuous infusion instead of intermittent injection of clotting factor, or the use of fibrin glue, should cut down overall need. The really important part of any surgical treatment programme is to prevent untoward bleeding in the first place. Inadequate control results in wound disruption, the possibility of infection, further bleeding, and the need for more clotting factor;
- he may not need clotting factor if he has factor VIII deficiency but does not have severe haemophilia A. Instead his response to desmopressin (DDAVP) should be known. Desmopressin is cheaper and safer than clotting factor transfusion;
- if he has a high-titre inhibitor and he bleeds he should not be moved around until things have settled down. Immobilization

of the affected area reduces the chance of rebleeding and further tissue disruption. This is especially important following a psoas or retroperitoneal bleed. Even the development of a bedsore because he has not been moved frequently may be preferable to another bleed in these areas;

- he should not ride a bike without wearing a safety helmet. Any activity that may result in head injury demands the appropriate protection;
- he should not be encouraged to lead a sedentary lifestyle. Lean, fit, and muscular people with haemophilia bleed less than fat, unfit, and flabby people with haemophilia!

Share the burden

Few severely affected people with haemophilia can manage on their own. They have to rely on family and friends for help when bleeding episodes and the pain of joint and muscle bleeds disrupt everyday life. But relatives have to lead their own lives and most have to work. Therefore it makes sense for affected families to befriend one another and share the care of their haemophilic members. The sharing of knowledge and equipment (not used needles, syringes, or giving sets of course!) helps, and a group of people have a more powerful voice to seek help from politicians, charities, and from the medical profession, than the man on his own. And, believe it or not, fund raising can be fun!

One of the initiatives being developed by the World Federation of Hemophilia is for the sharing of experience between centres in different parts of the world. It is thought that centres in developed and developing countries could be 'twinned' or 'paired' in the same way as many cities, so that they could help each other.

Befriend a physiotherapist!

This is probably the most important suggestion of all. It is quite remarkable how even the most severely affected person with haemophilia can keep going, despite intermittent bleeds and arthritis, providing he has the motivation to keep himself fit.

Much of the health of the major joints depends on muscle tone and power, and sometimes on the protection afforded by muscle bulk. Wasted, flabby muscles do not protect joints but increase the likelihood of a series of bleeds and worsening arthritis. A physiotherapist can do a great deal to improve the quality of life of someone with haemophilia provided she (or he) has the patient's commitment. Given motivation, people who have been confined to bed or a wheelchair for months or even years can be encouraged to learn to walk again—without blood product cover. All that is needed is the commitment to try, regular sessions, and mutual trust. The magic comes later!

Physiotherapists can help those with haemophilia in a number of ways, some of which are described in more detail in Chapter 8. They can teach how to:

- Take a pride in the body and enjoy working out with others. Healthy competition never does anyone any harm. Exercises alone at home can be so boring that it is easy to lose interest. Self-esteem and a pride in the body—body image—is fundamental to personal esprit. Teenage youngsters with haemophilia do not necessarily have to emulate the latest pop star. Experience suggests that many of them already look a good deal healthier!
- Control weight. Fat people put unnecessary pressure and strain on their bodies, and especially on the ankle, knee, and hip joints.
- Relax and stretch muscles and joints so that the body becomes more mobile and more able to counter stress. Previous contractures resulting from poorly treated muscle bleeds may be eased considerably by regular stretching exercises, especially if these are started early and built up gradually. Muscle spasm or stiffness is more likely to result in haemorrhage. Disciplines like yoga or hypnosis may also help relaxation.
- Start participation in sport. Cycling, rowing, and swimming are the best all-round sports for someone with haemophilia to try.
- Gradually build up muscle tone and strength, either generally or in a particular limb or group of muscles.
- Modify footwear to alter load-bearing on a target ankle, knee, or hip joint. This can be done by wedging or building up footwear to counteract leg shortening or pelvic tilt.
- Use shock-absorbing soles and heels in shoes in order to lessen loads on joints during walking or running.
- Restore joint range by graduated exercises, or by traction, or wedged plasters.
- Apply short-term, cheaply made plaster of Paris splints. The pain of chronic arthritis in a knee may be helped by wearing a full limb calliper for a while. It takes the pressure of body-weight off the injured joint, transmitting it instead directly from pelvis to heel.
- Make night splints which help prevent repeated bleeds in the same joint during sleep.
- Avoid overprotection by encouraging carefully controlled exercises once the acute symptoms of a bleed have started to settle.

- Become mobile quickly, by using exercises graduated from partial to full weight-bearing in the case of lower limb bleeds. This can often be achieved more quickly with the help of splints or, in the case of lower limb bleeds, calliper protection.
- Provide aids to mobility, for instance walking sticks, frames, or crutches.

Arrange regular follow-up

Encourage a local doctor or nurse to take an interest in the group. Only by monitoring the health of individuals over a period of time can the effects of any treatment be assessed, and recommendations for change be made. In addition, people with haemophilia are just as likely as anyone else to suffer from other diseases and disabilities. The sooner these are diagnosed and dealt with the less likely they are to cause problems with the underlying bleeding disorder. It is, for instance, sensible to check an adult's blood pressure and urine at least once a year. Hypertension and diabetes are easily controlled by medicines. Left untreated hypertension may result in cerebral bleeding which in the person with haemophilia can, of course, be catastrophic, and untreated diabetes can kill insidiously in months.

In developed countries nurses with experience in haemophilia care have proved to be invaluable members of centre teams. They often spearhead the care of affected families, provide the immediate treatment of the haemophilia, and share in the education of the wider community. A group that can befriend a nurse as well as a physiotherapist is well off indeed!

Join the appropriate local and national Haemophilia Society, and the World Federation of Hemophilia (WFH)

This is not just an extension of fellowship but a commitment to learning about how others tackle the problems of haemophilia world-wide. People living in remote areas will solve specific problems more easily and more quickly if they develop outside contacts with doctors and others involved in haemophilia care. The newsletters and meetings arranged by these organizations provide a way of making these contacts. Some national societies adopt centres in developing countries. For instance the UK Haemophilia Society has links with India—and the WFH (p. 346) sponsors a number of International Haemophilia Training Centres (IHTCs p. 264) and schemes for doctors and paramedical staff wanting to learn more about haemophilia. These schemes include workshops in areas where there is already a commitment to haemophilia care but the need for more information and guidance. A list of places with these centres is printed at the end of this chapter.

In 1990 Charles Carman, the WFH President, initiated a strategic plan, the main thrust of which was to try and bring help to the 80 per cent of people with haemophilia in the world who presently receive little or no treatment. The implementation of this, the Decade Plan, could be crucial if and when a genetic cure of haemophilia becomes a reality.

Encourage the development of local laboratory facilities

Specific intravenous treatment of the bleeding disorders cannot be carried out safely without laboratory backing. Many of the tests needed are easily learned and are simple and cheap to perform. This is especially so if reagents are made locally rather than bought on the commercial market. The purpose of the laboratory should be to make an initial diagnosis, and in particular to differentiate between haemophilia A and B, to have a go at assessing severity by assaying the level of factor VIII or IX, and to screen for clotting factor inhibitors. The WFH training scheme helps in the teaching of staff for, and the establishment of, these laboratories.

In planning a laboratory for the diagnosis of the bleeding disorders and the monitoring of their treatment it is very important to build in a system of quality control. This means that the results of the tests performed in the laboratory should be compared with the results of other laboratories, and with other tests using standard reagents at regular intervals.

Encourage local blood bank facilities

This is probably the hardest goal of all to achieve. Many blood banks in developing countries are so poorly staffed and equipped that they can only be expected to try to provide for the immediate needs of whole blood transfusion. Even the most rudimentary haemophilia care with fresh frozen plasma (for factors VIII and IX) or cryoprecipitate (for factor VIII) demands centrifugation and reliable deep-freeze facilities. Given these circumstances it is hardly surprising that the freeze-dried concentrates should be so attractive to both staff and patients. The vials can be prescribed off the shelf from a pharmacy just like any other drug. Provided they are kept in a cool place they do not require special storage facilities, they are sterile, contain a known amount of clotting factor, and they are easier to prepare and give than cryoprecipitate. They are, however, infinitely more expensive both because they are in short supply, and because they are difficult to manufacture. Despite this the commercial companies, forever on the look-out for new markets, will be keen to establish themselves anywhere that promises a long-term commitment to their products, and they may seal this commitment by offering initial inducements and price cuts.

Blood products are no different to other marketable goods on this level, and the same arguments apply. Without great care the potential for local, volunteer donor supply suffers and the World Health Organization (WHO) commitment to self-sufficiency in blood supply using volunteer donors fails. The development of genetically engineered products may eventually make the choice easier for people with haemophilia and their doctors in developing countries, but only after the considerable costs of research and development have been recouped.

My advice is to try to use concentrate for all the needs of the haemophilic community only *after* local facilities have been organized, and a safe fall-back position is in place should the outside supply dry up. Even then, because of the expense, buyers should try to restrict concentrate use to specific severe bleeds or complications. Surgery can be performed just as safely with cryoprecipitate as with concentrate.

Despite the world-wide shortage, from time to time companies have stocks of factor VIII or IX which are nearing, or past, their recommended expiry dates. These products are almost always still potent and safe enough for treatment for many more months, and should be given freely or at considerable discount to those in developing nations. However, potential recipients should beware of the 'dumping' of concentrates for reasons other than expiry dates. Thus products which have not been virally inactivated are unacceptable, as are those failing the quality control measures imposed by the authorities in developed countries.

Finally, always remember that blood product requirement might be reduced by the careful use of local haemostats, or simple devices like ice packs.

Look after the general health of affected families

Regular exercise, good diet and healthy living underpin the management of haemophilia. People should be encouraged to stop smoking and perhaps to curtail their alcohol intake. In themselves these measures alone will lead to significant improvements in the health of former cigarette smokers or those who relied too heavily on alcohol. And giving them up releases money for other things, including a better diet.

It is impossible to suggest a general all-round diet for everyone in a book of this kind because of the great variations between peoples, but the essentials are the same, and include adequate intake of carbohydrates, fats, proteins, vitamins, minerals, and clean water.

The World Health Organization (WHO) and the Food and Agriculture Organization (FAO), together with major charities and government departments advise on what is adequate for people of different ages in different countries. Pregnant women and growing children and adults engaged in heavy manual work have extra dietary requirements, which sometimes have to be met with supplements, for instance iron tablets in pregnancy. However, in general, supplements are not required and expensive proprietary multivitamin tablets and other preparations are a waste of money. Because muscle health is reliant on a good diet containing adequate protein, this is obviously of great importance to the people with haemophilia. Guidelines for healthy eating, especially for those with HIV infection, will be found in Chapter 14.

Get immunized

Many infections are preventable. Unimmunized people with haemophilia who survive the acute infection will be left with an additional handicap. Whenever immunization is available they and their families must take advantage of the offer and be vaccinated. At the very least everyone should be protected from polio, tetanus, and hepatitis B. A comprehensive schedule for immunization will be found on p. 269.

WFH International Hemophilia Training Centres (IHTCs)

Functions of these centres are three-fold:

- to train visiting physicians and technicians, particularly those from developing nations;
- to organize regional workshop programmes with lectures and demonstrations on the care of haemophilia;
- to send experts to developing countries who request help in the setting up of facilities for the treatment of haemophilia.

IHTCs are in:

- Argentina (Buenos Aires)
- Australia (Sydney)
- Austria (Vienna)
- Belgium (Leuven)
- Brazil (Rio de Janeiro)
- Costa Rica (San Jose)
- Finland (Helsinki)
- France (Paris)
- Italy (Milan)

- Japan (Tokyo)
- Sweden (Malmö)
- Switzerland (Basel)
- Thailand (Bangkok)
- UK (London, Oxford, Sheffield)
- USA (Chapel Hill, Los Angeles, New York, Philadelphia, Rochester, Worcester)

Full, updated details, may be obtained from the World Federation of Hemophilia, 1301 Greene Avenue, Suite 500, Montreal, Quebec, Canada, H3Z 2B2.

The World Federation of Hemophilia also publishes a directory and atlas of facilities available for people with haemophilia throughout the world.

PART FIVE

Families with haemophilia

14

Keeping healthy

This chapter contains background information on a variety of topics important to all those with haemophilia and their families.

Immunization against infectious disease

Having haemophilia confers no protection against infectious disease. Rather it raises the possibility of contact with infection, especially hepatitis. Despite the extremely good recent record of clotting factor concentrate safety, this safety can never be absolutely guaranteed. Immunization provides the growing child with an insurance policy.

Distinct from the slight risk of blood-borne infection are the other diseases that are now preventable by immunization. The person with haemophilia has enough to cope with. The additional burden of contracting an infection like polio or tetanus is devastating.

The recommended schedule of immunization in the United Kingdom is:

Immunization schedule in UK			
Age	Vaccine		
2 months	Diphtheria, tetanus, polio, Hib		
3 months	"	"	"	"
4 months	"	"	"	"
12–18 months	Measles, mumps, rubella		
4–5 years	Booster diphtheria, tetanus		
10–14 years	BCG. Girls only: rubella		
15–18 years	Booster tetanus, polio		
High-risk groups	Hepatitis A, hepatitis B, influenza		

Hib stands for *Haemophilus influenza* b, one of the causes of meningitis and blood poisoning in young children; routine immunization against Hib was introduced in 1992. BCG stands for Bacillus Calmette–Guérin, a vaccine against tuberculosis named after its originators.

People with haemophilia and those that treat them should be immunized against hepatitis B. Hepatitis A immunization is also recommended for those with haemophilia.

Immunization against infectious diseases for people with HIV infection is very important. However, there are some differences from the routine schedule. They are:

Immunization for people with HIV infection

- BCG should not be given because it can result in widespread dissemination of the injected organisms.
- Polio vaccine is sometimes recommended in its inactivated form which is injected, rather than in the usual active form which is taken by mouth.
- People with HIV infection excrete polio virus for longer than the usual 6 weeks after immunization. Careful hygiene is required to prevent cross-infection, especially if the contact is not immune to polio. For instance, hands should be washed after changing an immunized infant's nappies.
- People with HIV infection who come into contact with measles, chickenpox or zoster (shingles) need short-term protection with an immunoglobulin injection.
- Yellow fever vaccine is thought to be unsafe in those infected with HIV.

People intending to travel abroad may, of course, require further immunization specific to the countries to be visited. Enquiries should be made well in advance of the trip. Information is available from family doctors, haemophilia centre staff, and, in some countries, special travel clinics often run by the airlines.

Remember that the effect of vaccines wanes with time. Booster immunizations against polio and tetanus are wise precautions at 10-yearly intervals.

Further information about hepatitis will be found in Chapter 11.

The relief of pain

Pain is the major complaint of many people with haemophilia, and its management presents both the patient and his doctors with a problem. The patient rightly demands speedy relief from his distress; the doctor knows that all the drugs at his disposal carry side-effects. They may be harmful in people with bleeding disorders, they may impair concentration whilst at work or when driving a car, and the stronger drugs are known to cause dependence. Dependence is the desire to take the drug even when it is not really necessary. Strong dependence is the same as addiction, when the desire becomes overpowering.

Whilst there is no magical potion which will relieve all pain and leave the person taking it otherwise entirely normal, there are excellent products which, used sensibly, give great relief. The patient and the doctor must work together to find the right preparation, or combination of preparations, for the particular complaint.

Assessment of pain has to be a very individual undertaking. Everybody has a different pain threshold. In other words, everybody's perception of pain is different. What to one man is an ache, to another man is agony. Because of this it is difficult to accurately measure pain, and the doctor is reliant on his patient's own descriptions and feelings.

One of the difficulties met by patients with chronic joint damage or very frequent bleeds is the condition of *tolerance* to a drug. Tolerance is when the same dose of the drug produces a smaller effect each time it is taken. When this happens, it is often more sensible to ring the changes and switch to another drug than to keep pushing up the dose.

Doctors who are used to working with haemophilia know that demands for powerful drugs are not an excuse to 'search for kicks', but they may be concerned that the patient is putting too much reliance on the drug and not enough on himself. In the same way that potency of a drug may fall with tolerance, the patient may come to rely on a strong drug so much that he demands it for every pain no matter how mild.

Drugs are not, of course, the only answer to pain. Many people find relief with ice packs (p. 176) or warmth, splinting, or early exercise. Sometimes transcutaneous electrical nerve stimulation (TENS) may work (p. 177).

Overall the best treatment for pain is *the early injection of the relevant clotting factor.* Some people experience discomfort at the site of the bleed as the blood product is being given, but the pain then stops. Some find that once they have learnt to live within the limits imposed on them by their disorder their pains become less frequent and severe. The best solution for a particular individual is a matter of experiment. If all else fails, surgery may provide an acceptable answer. The removal of a troublesome joint can bring a rapid return to normal life after months or years of distress.

The following is a list of some drugs available for the relief of pain.

Analgesics (painkillers)
Aspirin (medical names: acetylsalicylic acid/ASA)

Medicines containing aspirin should never be taken by anyone with a bleeding tendency. Aspirin irritates the lining of the stomach and causes bleeding in people without haemophilia. In addition, it has a direct effect on the stickiness of blood platelets. So great is this effect that aspirin has been tested as an anticoagulant in people at risk from thrombosis and found to be an effective prophylaxis against heart attack in older men. There are hundreds of preparations containing aspirin on sale to the general public, and a list of many of them is given at the end of the book (Appendix B). Note that several of the preparations are advertised as cold cures, or for 'upset stomachs'!

Paracetamol (acetaminophen)

Paracetamol (acetaminophen) is marketed in Great Britain as 'Panadol' and a wide variety of other proprietary preparations. Paracetamol may be obtained without prescription as tablets or elixir for children. It has no effect on the stomach lining or on clotting and is the first analgesic drug of choice for people with bleeding disorders. It is very important to keep paracetamol preparations stored away from children. An overdose causes severe liver damage.

Codeine

Many codeine preparations are available without prescription. Some, including soluble products like Codis, contain aspirin and should not be used. Preparations containing codeine and paracetamol are safe.

Codeine phosphate is obtainable only on prescription. It causes constipation and is used for the treatment of diarrhoea and in cough medicines rather than as a painkiller.

Dihydrocodeine

Obtainable only on prescription in the UK, dihydrocodeine is more powerful than paracetamol. It is a useful painkiller after severe acute haemarthroses and deep muscle bleeds, has no effect on clotting, and does not cause constipation. It used to be marketed as DF118. A newer preparation is called DHC Continus.

Dextropropoxyphene

Dextropropoxyphene, marketed as Doloxene or, in combination with paracetamol, as Distalgesic, Co-proxamol, or Cosalgesic has about the same effect as codeine but may help some people.

Nefopam

Marketed as Acupam may help some people not responsive to paracetamol.

Morphine and similar compounds

These very potent analgesics are always used with great discretion as they are likely to cause addiction when taken too frequently. They are valuable for occasional use when pain is very severe, or after surgery. A side-effect of these drugs is depression of breathing, and their short-term use is carefully controlled for this reason. Drugs which may be taken by mouth in this group include pethidine, dextromoramide, (Palfium), dipipanone (Diconal), meptazinol (Meptid), methadone (Physeptone), pentazocine (Fortral), and phenazocine (Narphen). None is recommended for regular use. A morphine preparation, MST Continus, may be useful in the relief of very severe chronic pain. It is a slow relief preparation given twice a day by mouth.

Buprenorphine, marketed as Temgesic

This preparation is effective when taken under the tongue (sublingually). Like morphine, it can cause dependence and is a controlled drug in the United Kingdom.

Alcohol

Although excessive alcohol consumption is something to avoid, a tot of brandy or whisky will often do more for pain and worry than a handful of tablets. A 'wee dram' or a warm milk drink before retiring will provide comfort and help induce sleep. But beware of mixing alcohol with drugs!

Drugs which reduce joint inflammation

Some medicines found to be very effective in people with rheumatoid arthritis may help people with haemophilia who have chronic joint pain. They are particularly effective when taken regularly. These drugs, called collectively the non-steroidal anti-inflammatory agents, or NSAIAs, could be harmful to people with bleeding disorders because of their tendency to irritate the stomach and affect platelet function like aspirin. However, in the absence of suitable alternatives, some of them are prescribed on the understanding that if the patient suffers indigestion or evidence of more bleeding the drug should be stopped immediately. One, ibuprofen, is on sale in the United Kingdom without prescription. Other NSAIAs include naproxen, fenopren, and piroxacim.

There are many others. Indeed it seems that a new variant appears on the market every month. All this means is that no one drug is effective in everyone with chronic joint pain. Given the provisos on indigestion and more easy bruising or bleeding, together with proper supervision whilst the drugs are tried, it is certainly worth working with a doctor to ring the changes with these anti-inflammatory medicines to combat the pain of haemophilic arthritis.

NSAIAs are so-called because they do not contain steroids. Steroids are very powerful hormones. They are sometimes prescribed in an attempt to damp down the inflammatory response in the joints after acute bleeds, thus lessening the frequency and duration of pain. They have many side-effects and people taking them must be under careful medical supervision. The usual course for treatment of an inflamed joint, or for haematuria in haemophilia lasts between five and ten days. An initial high dosage is reduced day by day so that harmful side-effects do not occur. Patients on regular steroids must carry a special card as precautions have to be taken in the event of accident or need of surgery. The steroid preparations usually prescribed are called prednisolone or prednisone.

In some cases, a doctor may prescribe an anti-ulcer medication like cimetidine or ranitidine to take concurrently with an NSAIA or a steroid. In any case, all these drugs should be taken only after a meal or a glass of milk.

Sedatives and tranquillizers

Pain and anxiety often go hand in hand, and a combination of an analgesic and a sedative or tranquillizer will help some patients bear their pain more easily. In this group of drugs are chlormethiazole, diazepam, and promethazine which is sometimes prescribed to help children.

They are obtainable only on prescription in the United Kingdom, and should not be taken for long periods of time.

Pain, anxiety, and depression often disturb sleep. When this happens, life rapidly becomes intolerable, and the only answer is to try medication. However, before doing so simple precautions should be tried. The evening meal should not be taken immediately before retiring. Alcohol should be restricted in the evenings because too much, too late means poor sleep in the early hours. Spouses should discuss their problems in the early evening, and the cat should be in no doubt that darkness and peace should be synonymous.

If none of these measures works a number of very effective sleeping-pills are available. They include chloral hydrate, chlormethiazole, flunitrazepam, flurazepam, lormetazepam, nitrazepam, temazepam, triazolam, and triclofos.

If possible, courses of these drugs, and those used as tranquillizers during the day, should be short. This is because taken long-term they can result in a form of dependence, and some people find it very hard to stop taking them without experiencing unpleasant side-effects.

Precautions with all drugs

All drugs are dangerous if taken in excess. They must be kept well out of reach of children, preferably under lock and key or in a specially-designed safety cabinet. This applies even to relatively mild drugs, such as paracetamol, which causes severe liver damage if taken in excess.

All medicines should be kept in the containers in which they were supplied, and not put together in unlabelled bottles or boxes. Out-of-date and unwanted medicines should be returned to the chemist for disposal or destroyed and not hoarded.

If possible, drugs should not be kept on the bedside table. Quite apart from the risk to children, there is a chance that a second dose might be taken inadvertently when waking in a befuddled state during the night.

Remember that all medicines taken by mouth take some time to act, usually about half an hour. Another dose taken in this interval will not speed up the effect.

Many of the drugs mentioned above cause drowsiness and great care is needed when intending to drive a car or work with moving machinery or sharp tools. Alcohol in any form will increase this drowsiness and instability.

Always remember three things:

1. If your life is tolerable without medication don't worry, you are not peculiar!
2. If you feel you need the help of medication never feel frightened to ask the advice of your doctor.

3. Never be tempted to try someone else's medication. You may react to it adversely and end up in hospital.

Background information
Teeth

Teeth are formed in the jaws before birth, and the milk teeth usually start to erupt when the baby is about 6 months old. Children have 20 milk or deciduous teeth, all of which are usually through by the age of 2 years. The adult, permanent teeth replace the deciduous teeth gradually between the 6th and 25th years, the wisdom teeth erupting last. There are 32 permanent teeth.

Fig. 14.1 shows the structure of a typical tooth, with the inner pulp and the outer dentine and enamel. Teeth are living structures with nerves and blood vessels.

If food, particularly sugar, is left in contact with a tooth for a time it provides a perfect breeding ground for germs. These attack both the teeth, producing caries, and the gums, which bleed easily. Without the effective action of removing the caries and putting in fillings, teeth are slowly killed, a painful process. Caries is prevented if the diet contains small and regular amounts of fluoride. The easiest way of taking fluoride is in drinking water and in many areas it is added to the public water supply. At the concentration used it does no harm and it does prevent dental decay. Fluoride was in my family's drinking water and my three children, who were no better at cleaning their teeth than anyone else, had only one filling between them in their childhood.

Care of teeth

A few years ago someone with haemophilia needing dental extractions could expect to spend several weeks in hospital recovering from the

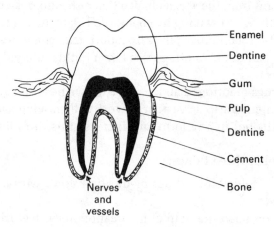

Fig. 14.1. Diagram of a tooth.

operation. Nowadays he is often back at school or work the next day. In spite of this improvement, the removal of teeth from a severely affected person is not without danger, and the procedure continues to demand the sort of back-up facilities described in Chapter 9. The development of inhibitors adds to the risk. Inhibitors *must* be excluded before dental blocks or extractions.

Baby teeth

The first teeth to be cut are usually the bottom front ones (incisors). The baby will want to bite on his new teeth and this should be encouraged with hard rusk-type foods. Soft biscuits and sweets stick between the teeth and form a perfect breeding ground for germs which destroy teeth and injure the gums. Few families can prevent their children eating sweets but they should try to withhold their introduction for as long as possible. When the inevitable happens and the child is given sweets by relatives or friends, parents should encourage the use of chocolate rather than sticky toffees as it will dissolve and wash away more easily. Some pamphlets on dental care recommend nuts as an alternative to sweets. Nuts should not be given to a preschool child because of the danger of inhalation into the air passages, with secondary chest infection. Crisps provide a safe alternative.

Dummies have come in for a lot of criticism from dentists in the past, and some parents hide them when they come to the clinic for fear of being told off! In fact, the ordinary dummy does no harm at all. Things to definitely avoid are dipping the dummy in sweetened juice, jam, or condensed milk, or giving the infant a 'comforter' in which a teat is attached to a plastic reservoir which contains juice. Nothing is more likely to wreck a child's first teeth, with consequent pain and sleepless nights for everyone. Comforters belong in the dustbin.

When the time comes, children with haemophilia lose their baby teeth without any bleeding problem. This is because baby teeth are gradually extruded from the gums giving plenty of time for healing to occur before they pop out. Very loose baby teeth do sometimes provoke a little bleeding before this happens, but no more than is usual in children without haemophilia. A quick tweak to separate the offending tooth from its owner, followed by a five-minute bite on a gauze pad (followed by an ice cream) is recommended.

Brushing the teeth and gums

It is a terrible reflection on how modern man looks after his health to realize that some people only buy a toothbrush to impress the landlady when they go on holiday. If the practice of brushing the teeth regularly is started early in life, the habit soon becomes as much a part of the everyday routine as eating. A baby's teeth should be cleaned regularly with a soft brush well before his first birthday, about a month after the first teeth have been cut, using water and not toothpaste at first. The child will be able to brush his own teeth properly when he is four-or five-years-old. Before this, he needs help.

A toothbrush with a small head and medium nylon bristles is recommended. Natural bristles do not dry readily and can harbour germs. If brushing hurts healthy gums, the brush is too hard. On the other hand, a brush that has become too soft and distorted should be thrown out. A new one costs very little if chosen wisely. Expensive brushes with exotic shapes have no advantages, although brushes that work between the teeth help to control plaque. The advertisements say that electrically operated brushes are 'recommended by dentists'. In fact, most of them are no more than expensive toys, especially the battery-operated varieties. Their only advantages are that they are more fun to use and easier for severely handicapped people with severe arthritis of their elbows or shoulders. Far more important than these gimmicks is the way the teeth are brushed and how often brushing is performed.

The brand of toothpaste chosen bears no relationship to the whiteness of teeth. All toothpaste seems to be inordinately expensive for what it is—a bland cream containing some form of abrasive, perhaps an antiseptic, and flavouring. Choose the cheapest brand! The only toothpastes which have a slight edge on rival brands are those containing fluoride.

In theory, the teeth should be brushed after every meal, but in practice most people brush after breakfast and before going to bed, and this is the habit to instil into children. The method of brushing is important. A quick dash across the teeth before tearing downstairs to watch the latest soap opera is useless—food debris still lurks between the teeth.

Fig. 14.2, taken from a leaflet prepared for the British General Dental Council, shows how effective brushing should be done. Bad technique may be checked with disclosing tablets or solution bought from the chemist. These colour the collections of bacteria with a special dye and thus identify areas where brushing should be improved. Areas between the teeth should be cleaned regularly with dental floss or a small interdental brush. If plaque is not removed, it combines with sugar from food to form acids which attack dental enamel and eventually make holes in the teeth. Plaque also leads to chronic gum disease, with loosening and loss of teeth. Only brushing, or the use of dental floss, can remove plaque; rinsing with proprietary washes is useless.

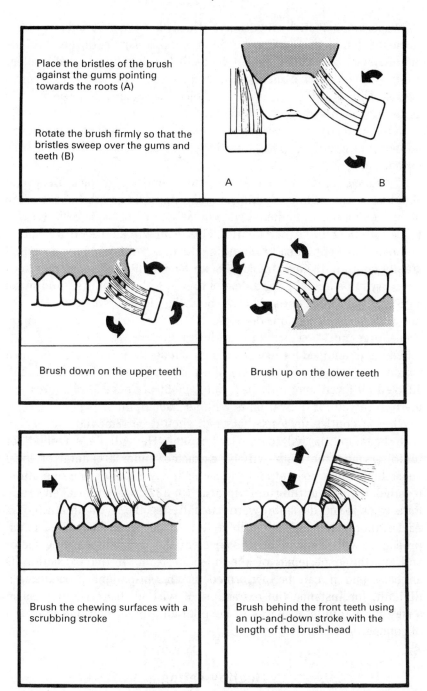

Fig. 14.2. How effective brushing should be done (from a leaflet prepared by the British General Dental Council).

Sweets and soft sticky foods have already been mentioned as a direct cause of dental decay. On the other hand, crisp foods help in the maintenance of healthy attractive teeth; apples, carrots, and celery are examples. After a meal, some adults use a toothpick to remove debris. This is safe in people with bleeding disorders provided the gums are healthy and the procedure is performed gently. Bleeding from the gums usually means that inflammation is present, a condition known as gingivitis, or periodontitis, which may be associated with pus formation.

Everyone should pay regular visits to a dentist, starting at the age of about two years. Affected children attending a haemophilia centre are often seen by the team dental surgeon and a dental hygienist who will work with the child's own dentist. This team will advise on all aspects of the child's dental health including the prescription of braces, which are not contraindicated in people with haemophilia. If the centre does not carry out a dental programme of this sort, the parents should make sure that the local dentist knows about the bleeding disorder. He will not want to give injections or extract teeth in his surgery; any operative procedures must be carried out in hospital.

Teeth are removed for reasons other than decay or infection, the most common being for overcrowding. With modern haemophilia treatment, affected children may now have the benefit of dental procedures formerly reserved for their friends without haemophilia.

When a dentist suggests that teeth should be removed the person with haemophilia will come to hospital. He will be screened for inhibitors and the teeth will be extracted under a general or local anaesthetic. One dose of clotting factor is given before the operation, together with an antifibrinolytic drug (p. 137), which is taken orally for a week to ten days afterwards, usually together with an antibiotic. An antifibrinolytic mouthwash may also be prescribed. The actual procedure will depend on the level of the deficient clotting factor and on the experience of the hospital team. If the extraction is an easy one it can be performed as an out-patient procedure. If difficult, for instance the extraction of wisdom teeth, it may mean a stay for a couple of nights in hospital and more than one injection of concentrate.

Healthy eating

A healthy diet is very important to someone with haemophilia. The maintenance of good muscles and stable, well-supported joints is dependent on regular exercise and quality nutrition. Poorly developed musculature and obesity add handicap to haemophilia. In

addition, a good diet is crucial for those infected with the human immunodeficiency virus (HIV).

General advice on healthy eating

People with haemophilia are just as likely to suffer from obesity and heart disease as anyone else. Therefore the rules for healthy eating are the same for all the family. No exceptions! The daily intake of most people in developed countries provides an excess of the energy, protein, vitamins, and minerals needed for health. We eat too much fat, sugar, and salt and not enough fibre and fresh fruit and vegetables. Therefore:

- Cut down on fats, especially the saturated fats of animal origin found in red meat, lard, and dairy products. Pastries and pies, cakes, chocolates, burgers, and sausages also contain animal fat. Use polyunsaturated fats instead. Find them in sunflower, corn or soya oils, margarines, nuts, and fish. Drink semi-skimmed milk and eat chicken in preference to red meat. Grill rather than fry foods. Remember that cutting down on fat will reduce the number of calories you are eating. In the case of animal fat it will also reduce cholesterol levels and protect the heart.

- Reducing fat intake often means less sugar too. Soft drinks, sweetened breakfast cereals, and, of course, sweets also contain sugar. Use an artificial sweetener for tea or coffee and choose calorie-reduced drinks.
- Use less salt in cooking and try to add less to your food once it is on the table.
- Go for high-fibre foods like unsweetened cereals, peas, beans, and lentils, wholemeal bread, and pasta.
- Enjoy fresh fruit, and fruit juices rich in vitamin C.

In the UK, the Health Education Authority publish a free, excellent *Guide to healthy eating*. It should be available from your local library or doctor's surgery.

Eating when you have HIV infection

HIV infection often leads to weight loss and muscle wasting. This adds to fatigue and depression, and makes it harder for people to keep active and warm. The suggestions given here have been adapted from those used by the staff of the Dietetic Department in the Royal Victoria Infirmary, Newcastle upon Tyne who have worked closely with my own staff in the haemophilia centre.

Healthy eating during illness

High-energy high-protein diets are recommended for people who are losing weight or who can only manage to eat a small quantity.

Your treatment or the effects of illness may decrease your appetite. So increasing the energy and protein content of your diet will help to make you feel stronger and not lose too much weight.

Your diet should always be balanced; this means including foods from each of the groups below:

Protein — Meat, fish, eggs, cheese, milk, pulses, nuts, beans, yoghurts;

Fats — Butter, margarine, lard, cooking oil;

Carbohydrates — Bread, cereals, rice, pasta, biscuits, cakes, potatoes, sugar, jams.

Boosting energy and protein can be done by:

1. adding manufactured food supplements;
2. adding everyday foods high in calories and protein;
3. making nourishing drinks.

Adding manufactured food supplements (Table 14.1)

Adding everyday foods high in energy and protein

Milk can be fortified with milk powder. Try to use about 2 pints of milk a day. Milk powder can be added to soup or milk puddings to boost the protein value.

Eggs are high in protein and can be added to mashed potato, milk puddings, and soups. Recent problems with salmonella in raw eggs

Table 14.1 Examples of products presently available on prescription in the UK

Product	How to use
Energy supplements	
Caloreen, Maxijul, Polycal, Polycose	Add 1 teaspoon to drinks, soups, milk puddings, or instant desserts
Polycal liquid	High-energy, flavoured drinks
Fresubin	A high-energy liquid to sip
Maxijul liquid, Hycal	As a syrup poured on fruit, or in recipes
Protein supplements	
Maxipro, Promod, Protifar, Casilan	Add to soups, milk puddings, casseroles, flan filling, fruit juice, etc.
Complete feeds	
Clinifeed, Nutrison Standard or Fibre Energy Plus, Nutrison, Fresenius, Fresubin, Ensure Plus, Jevity, Osmolite, Triosorbon,	Can be drunk if necessary to replace meals. Flavouring may need to be added
Supplementary drinks	
Fortimel, Fortisip Energy Plus, Fresubin, Formance.	Pleasant tasting between meal drink
Not available on prescription in the UK	
Build Up	Build Up and Complan are available in sweet and savoury flavours. They can be fortified with energy protein supplement, ice cream and milk shake flavouring; for example Crusha Syrup could be added

suggest that it may be better to use dried, pasteurized whole egg for this.

Grated cheese can also make a tasty high-protein, high-energy addition to mashed potato, scrambled egg, or cooked vegetables.

Cream is high in energy; add it to drinks, soups, puddings, breakfast cereals to increase calories. If you don't like cream try evaporated milk. Honey, jam, or marmalade can be spread on bread or added to porridge or milk puddings.

Add mayonnaise to sandwich fillings such as tuna or egg; it boosts calories and makes a tempting change. Peanut butter is packed with calories.

Butter or margarine can be melted on to vegetables, or perhaps spread a little thicker on bread and toast.

Nibbles between meals are a good idea. Keep foods like nuts, crisps, and fruit handy.

Nourishing drinks

These can be taken between meals or can replace a meal if you are unable to eat.

Home-made milk shakes are made in a liquidiser or with a hand blender. Suggested ingredients are milk, milk powder, ice cream, milk shake flavouring—Crusha Syrup or Nesquik—or tinned fruit.

Egg nog is made from milk, milk powder, egg, and a dash of sherry or brandy. Remember the warnings about the use of raw egg.

Special high-energy supplements (for instance Maxijul/Caloreen) can be mixed with these drinks. These have no taste and are, therefore, useful if you find other foods or drinks too sweet.

A sample menu The sample menu (Table 14.2) shows how small meals can be supplemented to increase their nutritional value. Taking drinks or snacks between meals is important because of the valuable energy and protein they provide.

Eating problems
Sore mouth

- use a mouthwash;
- drink through a straw;
- add ice to drinks,
- extremes of temperature may aggravate your mouth; if this happens avoid very hot and very cold foods;

Table 14.2 Sample menu

Breakfast	Porridge	—Make with fortified milk and add sugar or treacle or syrup
	Toast	—Add butter/margarine while still warm with jam, honey, or marmalade.
	Tea	—Add energy supplement
Mid am/pm	Milky coffee or milk shake	—Add energy supplement
	Biscuit	
Lunch	Soup	—Add milk powder or cream
	Tuna sandwich	—Add mayonnaise to boost energy
Evening meal	Meat	
	Vegetables	—Add butter
	Potatoes	—Add grated cheese, butter, or margarine
	Stewed fruit	—Add sugar or energy supplement
	Custard	—Make with fortified milk, stir in cream, add energy supplement
Bedtime	Hot chocolate	Make with fortified milk, add energy supplement

- avoid salty, spicy foods, or rough foods like toast and crisps;
- moisten foods with gravy or sauces.

Remember that in HIV infection a sore mouth can be due to thrush, which will respond quickly to treatment.

Dry mouth

- sip drinks; some people find mint tea refreshing;
- make iced lollies at home using fresh fruit juice;
- fortical or Hycal can be frozen in ice lolly containers;
- suck boiled sweets or peppermints.

Taste changes

A common complaint is one of an unusual metallic taste in the mouth. This is sometimes a result of your treatment and is usually only temporary. Concentrate on foods which you like the taste of and as your taste returns to normal try the foods you have been avoiding. Use more seasoning, herbs, and spices in your cooking. People often complain about the taste of meat—strong sauces such as curry, sweet and sour, or the addition of pickle or chutney may improve the flavour. Sharp tasting foods are refreshing and leave a pleasant taste in the mouth.

If you are too tired to cook, convenience foods, frozen meals, boil-in-the bag meals, and take-aways are perfectly acceptable. If you have a freezer you can buy or prepare meals in advance and freeze them. Suggestions of conveniently prepared foods are shown in Table 14.3.

Cooking by microwave is quicker than conventional cooking for many foods, but do follow the manufacturers' recommendations and the instructions on product labels.

Nausea

Avoid greasy, fatty, or fried foods. If the smell of cooking makes you feel sick, eat cold meals or food from the freezer which only needs to be warmed up. Let someone else do the cooking. Dry foods such as toast or crackers often help—try eating them first thing in the morning. Start off with light meals like thin soups or egg custards, and gradually build up your portion sizes and types of food.

Problems with chewing or swallowing

Try to make soft diets nutritious and appetizing. Some suggestions are shown in Table 14.3.

Table 14.3 Suggestions for conveniently prepared foods

Home-made soup or commercial soups can be fortified
Milk puddings
Sweet and savoury mousses
Flaked fish in sauce
Stewed or pureed fruit
Pasta dishes
Pancakes
Egg custard
Milk jelly
Cottage cheese, grated cheese
Porridge
Sandwiches with crusts removed
Creamed potatoes with a moist meat, fish, or egg dish

Remember that sherry is an appetite stimulant, and that all forms of alcohol are high in calories.

Sleep

There's nothing like a good night's sleep, especially if it is a rare event. Everyone has a different sleep pattern but, in general, we sleep for a shorter time as we get older, which seems unfair. The teenager emerging from his disordered room at midday has needed his sleep. People of my age can usually be found having a cup of tea in the kitchen in the early hours of the morning.

Haemophilia can result in disrupted sleep either because of the pain of a recent bleed or because of worry. The last thing doctors now advise is that the answer to insomnia is a pill. Unfortunately all the sleeping pills available may cause long-term dependence, and there are further problems with disordered sleep in coming off them. When stopping after a long course do it very gradually. Short courses prescribed for a specific reason are safe, but first there are some common sense measures you can take to try and get a good night's rest. They are:

- Do not eat a heavy meal late in the evening.
- Avoid too much alcohol in the evening.
- Retire to a comfortable, warm bed in a restful bedroom.
- Choose your partner carefully—there's not much you can do about snoring but a little romance does no end of good for one's sleep.
- If you do wake don't lie there ruminating. Read a book or get up and have a warm drink. Some people do the housework, but this tends not to be very popular with their spouses.
- Remember that regular exercise reduces stress. Physical tiredness is more likely to lead to a good sleep than mental fatigue.
- Relaxation techniques may help you. The physiotherapist at the haemophilia centre will give you advice.

The well-being of children in hospital

To severely affected children with haemophilia visits to hospital become an accepted, if unwelcome, part of life. Although most of these visits will nowadays be on an out-patient basis for regular follow-up or the occasional treatment, sometimes a child must be admitted to a ward. The admission may be for an operation or dental extraction, or for the more intensive treatment of a major bleed. When this happens, there are several ways in which his parents can help him settle down.

Very young children need the reassuring presence of a parent for as long as possible whilst in hospital. The ideal is for children under the age of five years to be admitted with their mothers; unfortunately some hospitals still do not have suitable accommodation for this. Most now allow unrestricted visiting. If family commitments or distance from home prevent the mother from being with her child for some time every day, she should try to arrange for another relative or friend to call and see him. The local branch of the Haemophilia Society might be able to help in arrangements for accommodation and visiting when the hospital stay is prolonged. Letters and postcards are very welcome whenever parents cannot visit.

Before a child comes into hospital, his mother should explain why the admission is necessary in terms suitable for his age. He should be told that the doctors and nurses are there to make him better, and that he will be able to come home as soon as he is well. If an operation is planned, his questions should be answered simply and truthfully, and he should know about the special magic sleep (anaesthetic) and told that when he wakes up the operation will be over.

He will want his favourite toys or books with him. Like Linus of Snoopy fame some children have a special piece of blanket they like to cuddle before sleep. This should also be taken, and the nurses told about it.

When leaving him after visiting, parents should tell him truthfully when they will be coming back. It is often a good idea to leave something (glove, handkerchief, purse) for him to look after while they are away. He will always cry when his parents leave, but the tears will be short-lived. His attention will soon be attracted by one of the nurses or playleaders.

Older children will have school, television, and computer games to keep them happy. If a long stay is expected, arrangements should be made for their own school books to be brought in and, whenever possible, for their own teachers to visit. Messages, cards, and group projects from their class are especially welcome. One such project I like is a T-shirt signed by everyone in indelible colours. A gift like this says 'get well soon and wear me', and is of especial importance to children who are very ill.

Sports

Children of all ages should be encouraged to participate in sports. By the time the boy with haemophilia has reached adolescence he should know which activities he enjoys. He should also be learning those activities which result in bleeds and, if these occur during a sport he has set his heart on, whether he can control them with prophylactic

Table 14.4 The top 10 sports recommended by doctors. All (100 per cent) of those replying put swimming at the top of this list; 94 per cent recommended cycling.

Sport	Per cent
Swimming	100
Table tennis	100
Walking	100
Fishing	99
Dance	98
Badminton	98
Sailing	98
Golf	96
Bowls	95
Cycling	94

factor VIII or IX beforehand. Regular, enjoyable sport results in feelings of well-being and fulfilment, which help counterbalance the unease and loneliness that haemophilia can provoke. Teamwork is another powerful weapon in dispelling ideas of isolation or handicap.

Recently, on behalf of the World Federation of Hemophilia, I asked colleagues for their views on sports most suited to people with severe haemophilia A or B. All were agreed that swimming, table tennis, and walking were to be encouraged, and many more sports were recommended. The 'top 10' sports are listed in Table 14.4.

Table 14.5 shows the least recommended sports, some of which carry a major threat of head or neck injury.

Altogether the questionnaire listed 69 sports. Bunjee jumping was not one of them, although I know three people with haemophilia who have tried it. It would terrify me, and is probably the quickest way of finding out if there are weak blood vessels in the brain. If you have haemophilia prophylaxis is advisable first!

Football (soccer) was well down the list, some doctors thinking that at a competitive level the risks of damage to the legs is too high. I do not share this view, preferring to let youngsters make up their own minds. What *is* essential for football, as for any sport, is that the usual protective clothing is *always* worn. In cricket, no self-respecting batsman (if there are any left) would dream of facing Merv Hughes without all the gear.

Hard hat and ankle boots when horse-riding, shin-guards when playing soccer, helmet and knee and elbow padding when skateboarding, a helmet when cycling, eye visor when playing squash, and shock-absorbing shoe and heel inserts when jogging are examples.

Essential parts of any sport are warm-ups and warm-downs. Not

Table 14.5 The least favoured sports. All
the doctors were opposed to boxing and 90
per cent thought skateboarding dangerous for
someone with severe haemophilia. (Shortly after
compiling this list I came across a champion at
karate. He has severe haemophilia, but hardly
ever bleeds because he is so fit and his joints
are so well supported by strong muscles.)

Sport	Per cent
Boxing	100
Rugby football	99
American football	99
Karate	94
Wrestling	93
Motorcycling	91
Judo	91
Hang-gliding	90
Hockey	90
Skateboarding	90

taking the time to loosen up before strenuous activity and to wind down afterwards is to set yourself up for injury. Muscles that have not been gradually stretched before exercise are far more likely to be 'pulled' than muscles treated to a warm-up session. And once an injury is provoked the only way full recovery can be achieved is with rest, often of several weeks duration. Your physiotherapist or trainer will design warm-up and warm-down programmes with you.

If you are unlucky enough to have a sports injury get early, prompt treatment for it. A soft tissue injury, for instance a strained or ruptured ligament, may be masked initially by a muscle bleed. Therefore do not be fooled by thinking that results of injury are *solely* linked to your haemophilia.

Some people keep themselves in trim by practising a regular programme of exercises at home. In order to pursue this dream, they invest in the sort of expensive (and often poorly built) equipment which always looks good in the colour supplements. Indeed advertisers continue to encourage everyone to leap around every day so that they look like Jane Fonda (which presents us men with a bit of a problem). The wife of one of my friends bought him an exercise bike, so he bought her an armchair. The bike has yet to be used but they both like the armchair!

It takes a very special commitment to exercise regularly on one's own. Far better to join a decent gym and get a kick by exercising with others. Look for a gym frequented by the local harriers or other sportsmen and women, and beware of glitzy places with floor to ceiling equipment and excessive fees. Your haemophilia centre physiotherapist may be able to advise you, or ring round the local athletic clubs for guidance. Once you've started try to make a commitment to go regularly—the hardest part of any sport is getting out of that chair! When you do exercise regularly you will get that 'buzz' that gives you such a lift, a lift that improves you both physically and mentally, and helps you shrug off many of the potential setbacks of haemophilic life. It also helps those with haemophilia to develop a positive body image. This is especially important if any degree of chronic arthritis is present as a result of recurrent bleeding.

Some forms of sports activity should be encouraged in everybody with haemophilia, but children should not be pushed—rather they should be given the opportunity to develop whatever pastime they want. They will soon learn which activities are in keeping with their disorder.

Practised regularly, a sport aids mental concentration, exercises and tones up muscles, and increases skill, co-ordination and, stamina. It fosters team spirit and comradeship and can give the most severely affected person a sense of personal achievement which may be lacking in his everyday life. Some people may think sport involves unnecessary risk, but risk-taking is not a neurotic manifestation of haemophilia; rather it is a necessary component of full emotional development. The essence of a sport is the learning of skills which reduce the risk. The man who goes climbing or sailing without the proper equipment or training is not only foolhardy but will endanger the lives of others as well as his own—an act which is indefensible. Provided the person with haemophilia is not foolhardy, he will be as safe as anyone else.

Youth clubs, scouting organizations, camps, and holidays

The more the severely affected person gets out of the house and keeps both mentally and physically active the better. Many clubs and organizations have special membership for disabled people and welcome their participation. At summer camps, arrangements can usually be easily made for medical cover. Indeed, some societies run special camps for youngsters with haemophilia. A world-wide list is being prepared by the World Federation of Hemophilia. Exchange visits between countries can often be arranged. Affected children should not be barred from school or club trips or picnics. As children mature holidays away from parents foster independence and should be encouraged.

The question of travel away from home raises problems of both treatment and insurance, especially when living with HIV infection as well as haemophilia has to be faced. There is now a world-wide network

of haemophilia centres which will provide the correct treatment should it be needed on holiday. Details can be found in *Passport*, a publication of the World Federation of Hemophilia.

When planning a journey, the family should find out about local facilities, either by enquiring at their own hospital or through the national Haemophilia Society, which will be affiliated to the World Federation of Hemophilia. Depending on individual needs and length of time away from home, they may wish to take a basic kit of medical equipment, including splinting material (plaster of Paris bandages or stiff plastic sheet), crêpe bandages, padding, and Band-aids, and, if available and applicable to the disorder, vials of special concentrate and infusion kits. The concentrate may be carried in an insulated picnic box to keep it cool, and can be stored in a cool place or a refrigerator during the holiday.

Also carried should be information about the disorder, which must include an accurate diagnosis, level of affected factor, details of inhibitors, blood group, and warnings of allergy or sensitivity to drugs. The centre doctor will be able to give this information.

Documentation may be required by Customs in order to clear the medicines, concentrate, or needles being carried. In addition to a letter for the Customs and the other documentation, I give HIV infected people a personal letter of introduction to medical colleagues in the country they are visiting. This letter contains the confidential details of their infection and its treatment. It is kept separate from their other papers and only used when, and if, it becomes necessary.

Sometimes travel company insurance does not cover people with bleeding disorders. People with haemophilia should never be misled by *carte blanche* statements from holiday operators. Careful reading of the small print will sometimes reveal an exclusion clause to the effect that the policy does not cover medical expenses for chronic disorders, or HIV and AIDS. It is particularly important to explore the question of insurance well in advance in the case of an affected child going on a school trip—the education authority may refuse a place if the child is excluded from party cover because of his disorder. Insurance can usually be arranged with one of the large companies if a detailed medical report and opinion on fitness to travel are provided by the patient's specialist, although this may mean revealing HIV infection. Haemophilia societies often help with insurance problems, or the hospital specialist may know of a reputable broker who is willing to give advice.

When on holiday, the temptation to do everything at once on the first few days should be resisted. Bleeds will be provoked in the boy (or the adult!) who spends hours in the swimming pool on the first day when he is not used to the activity, and the rest of the holiday may be a disaster.

One form of holiday which parents sometimes ask about is the advisability of a pilgrimage to a religious centre; in Europe, trips to Lourdes are popular. There is, of course, no harm in this, and there may be much benefit in terms of emotional acceptance if not in physical recovery. The family must make the decision in counsel with their religious adviser, but the same strictures with regard to treatment and insurance that apply to a conventional holiday are applicable.

Older people

Much of what has been said about sports and activities in youth applies to the older person with haemophilia. The quality of his life will be improved if he keeps active, and the pain and restriction of movement associated with chronic arthropathy will be lessened. It is particularly important to avoid the hazards of obesity. The fat, overweight person is not only more liable to accident, but is constantly putting a great strain on his joints which will wear out more quickly.

Although physical activity has received emphasis in this chapter, it is, of course, equally important for the person with haemophilia to keep active mentally as he grows older. The enormous pleasure to be gained from the arts need hardly be mentioned, and haemophilia societies often run postal clubs for activities like stamp collecting, photography, scrabble, computer games, and chess.

Remember that many older people with chronic disorders are particularly good at helping others. They have been 'through the mill' and known the frustrations and difficulties encountered by youngsters. The voluntary services, which include the haemophilia societies, are always on the look-out for wise counsellors and workers.

15

Planning a family

Since the first edition of this book in 1974 the sexual revolution has resulted in a more open and honest attitude to sexual behaviour, with free discussion allowed in many countries where sex was previously a taboo subject. The speed of this welcome change has been remarkable. However, it has left an assumption that everyone feels comfortable about sex, and that all young people have been armed with accurate knowledge by the time they become sexually active. That this is not the case is demonstrated by continuing unwanted pregnancies and sexually transmitted diseases. So, although the love between two people is a very private and personal part of the pattern of their lives, when ignorance about sex and family planning intrude they may bring a lot of misery and worry. This is particularly true when couples know that their children may inherit a disorder like haemophilia.

This chapter has been written to provide people with knowledge about how haemophilia can colour human sexuality and how easy it is to put things right. It also covers the broad questions relating to family planning and advances in our understanding of early life. In reading it, it may be difficult to remember that sex should be great fun and that it is meant to enrich our lives! Nowadays young people need the reassurance that sex should be a happy sharing of their most intimate experiences with someone they love. The doom and gloom of much anti-AIDS advertising can do them a disservice. Armed with knowledge and the tenet of caring for others they will not go far wrong.

Background information
The reproductive system
In very early life in the womb the sexual organs of boys and girls look exactly the same. The organs become different under the influence of a hormone produced in response to instructions carried on a sex chromosome. Later in life this common beginning may be seen in the genitalia of both sexes.

Boy and man

The visible sexual organs of the male are the penis and the scrotum, which contains the testicles (or testes). In childhood the testicles are small and at times may be hidden in small channels above the scrotum. Then, under the influence of a chemical messenger, or hormone, from a gland at the base of the brain, puberty begins. The testicles grow bigger, the penis enlarges, hair appears on the face and pubis and in the armpits, and the voice breaks. Puberty also heralds a spurt in the growth of bones and muscles, strength increases, and joints become more stable.

Adult testicles have two functions. The first is to maintain the flow of the male hormone testosterone into the blood, and the second is to produce the sperm needed for fertilization.

A tube, the vas, runs from each testicle up into the abdomen to the base of the bladder (Fig. 15.1). Here the tubes from each side join up and enter the urethra which runs from the bladder through the penis. The tubes join in the prostate and it is this gland, together with some smaller glands, which produce much of the fluid (semen) in which sperm are ejaculated from the body. The urethra opens at the tip of the penis and through it pass either urine or semen; a valve mechanism prevents both being passed together.

At birth the end of the penis is covered by a fold of skin—the foreskin, or prepuce, which is sometimes removed early in life for religious, personal, or medical reasons; this is the operation of circumcision.

Girl and woman

The visible sexual organs of the girl, the vulva, are the labia, or lips, which surround the cleft which runs between the legs; in the male these lips are fused together to form the scrotum. Towards the front of the cleft is a small knob of very sensitive tissue called the clitoris. This structure is equivalent to the penis and like the penis becomes engorged with blood and erect. Behind the clitoris (Fig. 15.2) lies the opening to the tube from the bladder, the urethra, and behind this is the opening to the muscular tube called the vagina. The vagina serves both as recipient to the penis during intercourse, and as the birth canal for the baby. In young girls the lower end of the vagina is partially closed by a membrane called the hymen or 'maidenhead', which is stretched or broken during early intercourse. Within the vulva and vagina are a number of small glands which produce a lubricant.

At its upper end the vagina meets the neck of the womb, or uterus (Fig. 15.3). The part of the womb which projects into the vagina is named the cervix, and it is from here that a small scrape or 'smear' is taken by doctors looking for the early changes that sometimes herald cancer, which if caught this early can be completely cured. A small channel runs through the cervix into the hollow of the womb, which is a muscular organ about the size and shape of a pear. Two tubes enter the womb, one from each side. At the other end these tubes open into the cavity of the abdomen near two walnut-like structures, the ovaries. The ovaries are the equivalent to the testicles in the male, and they too play a dual role; they produce female hormones, and they release the eggs (ova) after puberty.

Puberty in the girl results in changes similar to those in the boy. Hair grows over the pubis and in the armpits, and the skin becomes more greasy and liable to acne. The girl begins to grow more quickly, but she also becomes more rounded as fat is laid down over her hips. Her breasts develop and her genitalia enlarge. In both sexes the timing of puberty is very variable, but girls usually mature earlier than boys.

The main sign of physical maturity in the girl is the start of one of the rhythms of life we call menstruation.

Menstruation Menstruation is nature's way of expelling unwanted material from the womb periodically, hence the term 'period'. Within the womb is a lining of specialized cells. At the beginning of every 4 weeks these cells start to build a comfortable bed for the reception of an egg which, if it is fertilized, will become a baby (Fig. 15.4). After 2 weeks or so this bed is ready and the egg is discharged from the ovary. If it is not fertilized by a sperm, the egg does not embed in the thick lining, and after another interval of about 10 days the bed begins to break down. The lining detaches from the wall of the womb and a little bleeding occurs. It is this lining and the accompanying blood which forms the period; it usually takes about 5 days for all the lining to be rejected and for the bleeding to stop. The process then starts again in readiness for the next egg. The control of this mechanism is from hormones produced by the ovaries and other glands.

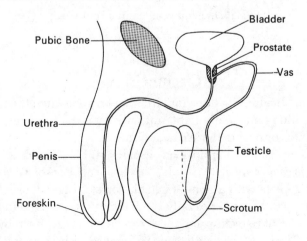

Fig. 15.1. Male anatomy.

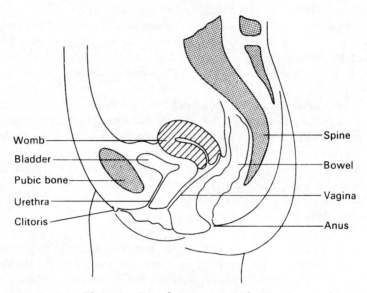

Fig. 15.2. Female anatomy (side view).

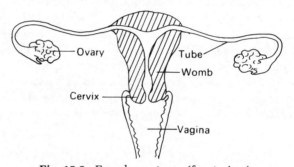

Fig. 15.3. Female anatomy (front view).

Circumcision

The only medical indications for circumcision are when the foreskin of an older child is unusually tight and will not retract, or when periodic inflammation obstructs the passage of urine.

In the baby the foreskin does not usually pull back, and although full retraction is easy in most 3-year-olds this is not always so. If a boy reaches the age of 7 years with a non-retractable foreskin the advice of a doctor should be sought. Sometimes only small adhesions between the skin and the penis prevent retraction and these can be painlessly freed by a simple procedure. The foreskin should never be forced back by a parent, but gentle easing at bathtime will do no harm over the age of 2–3 years.

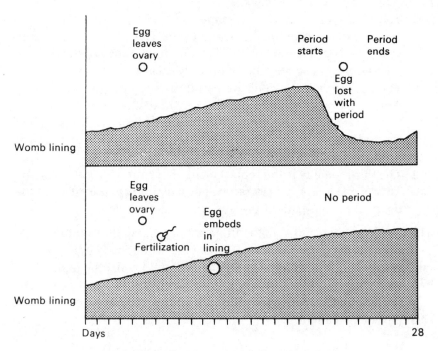

Fig. 15.4. Menstruation and pregnancy.

Families with a history of a bleeding disorder should always warn the doctor concerned of this, and where circumcision is indicated on religious grounds the appropriate authority should be told about the bleeding disorder.

Heavy periods

Girls with von Willebrand disease, or an autosomal recessive factor or platelet disorder, and occasionally haemophilia or Christmas disease carriers, will sometimes experience unduly heavy and prolonged periods.

The first periods in any girl are often irregular and may be either scanty or heavy. As the pattern of menstruation becomes established so the periods become more regular and of about 5 days' duration, with the maximum loss in the first 2 or 3 days. Mother should keep an eye on her daughter during these early years of womanhood. Heavy periods may be concealed by shy or frightened youngsters, and many months of unnecessary worry may precede an effective remedy. There is no reason why internal tampons should not be used by a girl with a bleeding disorder.

There are three medical remedies for heavy or prolonged periods:

1. The first is the use of a hormone pill, like that used for contraception, to regulate the menstrual cycle. The preparation usually prescribed is a combined pill containing a low oestrogen (30 micrograms) dose, together with progestogen.
2. The second is the use of the antifibrinolytic drug Cyklokapron. Antifibrinolytics prevent the breakdown of clots and may have a very good effect in reducing both the blood loss and the length of the periods; they are taken from when the period is established until the loss has ceased.
3. The third remedy is the replacement of the missing clotting factor. This is needed only very rarely, but when it is a quick trip to the hospital for out-patient therapy solves the problem.

The doctor in charge of a girl with heavy periods will take blood tests to check that she is not becoming anaemic. If she is, a course of iron tablets will soon put things right. Women should resist the temptation to buy expensive potions or tonics from the chemist. The doctor will know both when a medicine is needed, and which one is the best for each particular patient.

Sexual activity

The bleeding disorders have no direct effect on sexual activity. Only very rarely is a bleed likely to be associated with masturbation or intercourse, and then only in the most severely affected person. When a bleed does occur it is likely to be confined to minor bruising of the penis, but a bleed into the psoas or iliacus muscle may occur (p. 100). As a psoas bleed takes time to resolve, it is wise to refrain from intercourse for a week or so after restoration to full mobility after the bleed.

Young women with inherited clotting disorders need not worry that first intercourse will be disrupted by excessive or prolonged bleeding. When bleeding does occur it almost invariably stops quickly without medication.

Indirect effects are sometimes associated with the fear of an unwanted pregnancy, or with changes in a couple's usual sexual practice to provide an outlet for the frustrations of chronic pain or the strain of bringing up a child with severe haemophilia. In these, as in all sexual problems, people should try not to let shyness prevent them seeking the advice of their family doctor or hospital specialist.

Premarital advice

An unwanted child with a bleeding disorder has two handicaps instead of one and is perhaps less likely to be accepted for adoption. People with bleeding disorders or carrier states should bear this in mind before they enter into sexual relationships outside marriage. It is only fair that they take scrupulous contraceptive measures to avoid the possibility of pregnancy and the risk of an affected child. Before marriage both partners should know about the bleeding disorder and the risks to their children.

The young couple should have a chat with a trusted doctor and discuss the problems openly. If the doctor is not a specialist himself he will be able to suggest the names of experts in carrier detection and genetic counselling, usually at a haemophilia centre. Initially the couple should be seen alone, but they might like the facts to be explained to interested relatives on another occasion as well. Grandparents and other inquisitive relatives can easily get hold of the wrong ideas!

Family planning

Effective contraception is of especial importance to families with hereditary bleeding disorders. Parents will wish to limit the number of children who may be affected, and fear of pregnancy may bring strain to marital relations.

There are many methods of contraception. The one chosen by a particular couple will be as much a result of personal preference and perhaps religious conviction, as the result of purely medical choice. Some methods must always be prescribed or supervised by a doctor, and in comparison with the others these are by far the most effective ways to avoid pregnancy. The remainder of the chapter is concerned with the different methods and with sterilization and abortion. *With the exception of the condom none of the methods described here offer any protection against AIDS* if an uninfected partner has intercourse with an infected person.

Contraceptive methods
Withdrawal (coitus interruptus)

In ordinary language this is often referred to as 'being careful', but as a method it is not always careful enough. It is a very ancient and widely practised form of birth control and most couples have used it at one time or another, usually in early marriage. For some people it is the only acceptable method and is regarded as part of normal intercourse.

Studies in many countries have shown withdrawal to be fairly effective and without harmful effects. However, for the couple who definitely do not want another child this method cannot be recommended. Coitus interruptus means the withdrawal of the penis before the ejaculation of semen, and this is obviously not foolproof. Active sperm may be introduced into the vagina before climax, and in the circumstances of a second honeymoon or after several whiskies the desire to complete the act may become more pressing than the desire to avoid pregnancy.

The condom (sheath)

This is the most widely used birth-control device. In most countries condoms are easily obtainable, often in a bewildering variety of forms and colours, and with a remarkable number of names. In many countries condoms are officially tested for reliability; in the United Kingdom those that pass testing carry the British Standards Institute symbol on the pack. Provided the condom is used correctly it is an effective method of birth control. However, because of the very occasional risk of mechanical failure it is wise to use a chemical contraceptive in conjunction with the condom when there are absolute reasons to avoid pregnancy.

Condoms are made from thin rubber and are worn over the penis during intercourse. They provide a barrier between the penis and ejaculated semen, and the vagina. Their use is therefore twofold. Sperm is prevented from entering the woman and therefore pregnancy is avoided, and sexually transmitted diseases are unlikely to pass from an infected to an uninfected partner.

If the condom is used properly at every episode of intercourse it gives reliable contraception, about 3 pregnancies occurring in every 100 couples using the method regularly. Instructions for proper use are printed on every packet of condoms but are difficult to read in the dark! Therefore people becoming sexually active, both boys and girls, should learn the rules first.

1. Put the condom on before sex when the penis is erect. The fluid ejected from the penis before full ejaculation contains sperm, and, if this comes into contact with the vulva, pregnancy can occur without penetration of the vagina.
2. Gently squeeze the last centimetre of the closed end of the condom between finger and thumb so that it can receive the semen.
3. Unroll to cover the whole shaft of the erect penis.
4. After coming (ejaculation) hold the condom in place to prevent semen spilling out as the penis is withdrawn from the vagina.
5. Use a new condom each time you have sex.
6. Dispose of the condom by wrapping it well in toilet tissue and putting it in the bin or down the lavatory, or by incineration.

If a lubricating or contraceptive jelly or cream is used with a condom it should be water rather than oil based. This is because oil-based preparations like Vaseline can damage the condom and make it unsafe.

Recently a female condom has been marketed. Its effectiveness relies on the positioning in the vagina of an inner and an outer flexible ring contained within the plastic or latex sheath. This can be tricky at first and needs practice to get it right.

Chemical contraceptives

Like condoms, these come in many varieties. The types available include pessaries, jellies, creams and foams which contain chemicals to kill sperm. Those containing the chemical nonoxinol-9 are recommended, because it is both a spermicide and may help protect against sexually transmitted disease. Chemical contraceptives are not effective on their own, and they should always be used with a condom, cap, or diaphragm. Whichever type is used it should be placed high in the vagina shortly before intercourse. Repeated intercourse requires repeated applications.

Diaphragms and caps

These are devices which must be fitted for the individual woman with the help of a doctor or trained family planning worker. Diaphragms and caps work by acting as barriers within the vagina, blocking the passage of sperm to the womb, and can be inserted up to an hour before intercourse begins. The diaphragm is placed to lie diagonally across the vagina, its lower edge being retained in position by the pubic bone. The cap must be carefully placed over the neck of the womb itself. A chemical contraceptive should be used with both devices.

There are certain medical contraindications to the use of both diaphragms and caps, but when fitted properly and used carefully they provide an effective method of contraception. They give little protection against AIDS because there is no barrier between the partners except that over the cervix.

The safe period, or rhythm method

The word 'safe' in the name of this method is misleading, and the method is not recommended to those whose future children are at risk unless they are prepared to follow the procedures required very carefully indeed. The method depends on accurately forecasting the time of ovulation, that is when the egg leaves the ovary on its journey to the womb. The couple must abstain from intercourse during the 10 days around this fertile time.

Catholic families are, of course, in difficulty here. Many now follow their own consciences and choose an alternative acceptable to them. Priests are usually sympathetic to the problems involved, and worried couples should seek their help and advice, or that of the Catholic Marriage Advisory Council. If there is no alternative, then users should realize that to be in any way effective the rhythm method, which is also called natural family planning (NFP), must be applied rigorously. The method will not work if periods are irregular, in the months after the birth of a baby, and at the time of the 'change of life' (menopause). It is most likely to fail during holidays!

The pill

There is no doubt that oral contraceptives provide the most effective contraceptive method. Taken regularly the pill is better than any other method apart from sterilization, about one pregnancy occurring for every 200 women using a pill in a year. Remember that the pill gives no protection against AIDS.

A tremendous amount of work has been put into the consideration of the safety of the pill all over the world. Unfortunately the newspapers, in their quest for dramatic headlines, have highlighted the rare harmful effects of this form of contraception, with the result that many women have become unnecessarily frightened about its use.

The pill works by altering the levels of specific sex hormones in the body. Normally these hormones instruct the ovaries and the lining of the womb to carry out the monthly cycle of ovulation and menstruation. By altering the hormone levels artificially, the instructions to the ovaries are also altered and they will not release their eggs. A pill taken continuously (every day of the cycle) will also stop menstruation, and this frequency of dosage is sometimes used by athletes, when a period might interfere with their performance. Usually, however, the active hormones are only taken for 21 days, starting 5 days from the beginning of a period. Menstruation occurs soon after the last pill has been taken because the body hormones are no longer under the influence of the drug. The pill is then restarted at a specified time (usually 7 days). The timing is calculated from a normal period so that the days when ovulation is likely to occur are safely covered.

Some medicines, especially antibiotics, can make the pill less effective for a while. Check with your doctor, and if in any doubt either abstain from sex while you are taking them or use a barrier method as well.

If a pill is missed the family planning organizations give the following advice (shown in the box) to women.

- If you forget a pill, take it as soon as you remember and carry on with the next pill at the right time. If the pill was more than 3 hours overdue you are not protected. Continue normal pill-taking but you must also use another method, such as the sheath, for the next 7 days.

- If you have vomiting or very severe diarrhoea the pill may not work. Continue to take it, but you may not be protected from the first day of vomiting or diarrhoea. Use another method, such as the sheath, for any intercourse during the stomach upset and for the next 7 days.

- Those taking progestogen-only pills are also unprotected after 3 hours being overdue, and are advised to use another method for 7 days whilst continuing normal pill-taking.

Several types of pill are on the market and the doctor and patient will choose the one most suited to the particular circumstances. Full instructions are always enclosed in the packet. Serious side-effects are extremely rare, but a few women complain of an increase in weight, headache, depression, fullness in the breasts, or irritability. Most of these and other symptoms depend on the different hormone preparations and will disappear or become less of an annoyance by changing to a different sort of tablet.

Women wanting to start the pill should see a doctor who will take a history of their health and that of the family, and give them a short physical examination which includes taking their blood pressure. Similar checks should then be carried out at least once a year.

Side-effects of the pill On the positive side, apart from the fact that the pill remains the most effective form of contraception, are its beneficial effects. It makes menstruation easier for many women and is associated with a lower occurrence of pelvic inflammation and perhaps some forms of arthritis. More importantly we know that it has a protective effect against cancer of the ovary and cancer of the lining of the womb, both growths which are difficult to detect early.

On the negative side are the reports linking the pill with certain cancers and with thrombosis. Women with a history of abnormality in their blood circulatory system should not take the pill. Thus previous thrombosis, phlebitis, heart disease, high blood pressure, or hyperlipidaemia (high levels of blood fats, usually associated with a positive family history of abnormality) are strict contraindications to the pill. Other contraindications are cancer of the breast or womb, and the metabolic disease porphyria in which oestrogen or progesterone can precipitate a crisis. There are a number of other conditions when the pill should be used with care, after considering other, potentially safer methods of contraception. They include age (over 35 years), obesity, epilepsy, sugar diabetes, asthma, migraine, and severe varicose veins. Cigarette smoking puts women on the pill at additional risk of thrombosis.

Normal body function takes over in most women within 3 months of stopping the pill, with a return to regular menstruation and fertility. Taking the pill does not damage a future baby.

The morning-after pill

There is a 1 in 30 chance of becoming pregnant if unprotected intercourse occurs during the fertile days of the menstrual cycle. If this happens, continuation of the pregnancy can probably be avoided if two tablets of a combined contraceptive pill are taken within 72 hours of coitus. The pills recommended (2 tablets of Ovran or Schering PC4) should be repeated at 12 hours. Manufacturers recommend that these pills should not be taken if a period is overdue or more than 72 hours has elapsed. They also recommend that the woman should see her doctor three weeks after using this treatment.

Recently a new product called mifepristone (Mifegyne; RU 486) has been licensed for use in some countries. Dubbed an 'abortion pill' by the media, mifepristone counters the action of progestogerone. In very early pregnancy this leads to breakdown in the lining of the womb provoking bleeding and loss of the fertilized egg.

Depo-Provera

This is a synthetic progestogen which is injected every 3 months. Like the oral pill the hormone in Depo-Provera interferes with the menstrual cycle and acts as a contraceptive. Although approved for use in many countries and by the World Health Organization, debate continues about the preparation, mainly because any side-effects must be put up with for a longer time than is the case with oral hormones. As far as the bleeding disorders are concerned, the main drawback of Depo-Provera is that some women have rather heavy bleeding during their disrupted cycle. For this reason, and because it must be given by intramuscular injection, it should not be used in those with low clotting factor levels or von Willebrand disease.

Intra-uterine contraceptive devices (IUCDs)

Made of plastic, and sometimes incorporating copper wire or hormones, IUCDs are inserted easily, at least in women who have had a baby, into the womb through the cervix. It is thought that they act by preventing the acceptance of a fertilized egg by the womb. The menstrual cycle itself is unaffected but, as heavy bleeding may follow insertion or occur between periods, the method is not recommended to women with low clotting factor levels. For most others it provides an acceptable and safe method of contraception, the chances of pregnancy being less than with any other method apart from the pill. The failure rate is about two pregnancies for every 100 women using IUCDs a year.

Other methods

New ways of preventing conception are being researched. They include post-coital hormonal preparations and agents which act against sperm— the 'male pill', as well as variations on the barrier methods mentioned earlier in the chapter. Among these are sponges impregnated with sper-micide, new cervical caps and devices made to fit in the cervical canal. New ways of timing ovulation are also being advocated. These include the Billings method, which depends on learning how to recognize the natural changes in cervical secretions and is of relevance to couples practising natural family planning. Easier to use but more expensive are a range of ovulation prediction kits available from pharmacies.

Douching and the insertion of tampons or ordinary sponges into the vagina are not effective as contraceptives and may be dangerous because they can introduce infection or possibly the toxic shock syndrome.

Some old wives' tales are still heard and need to be discussed. The attainment of orgasm by a woman has nothing to do with the chances of conception. Nor do the phases of the moon. Breast feeding lowers a mother's fertility for a time but does not make her immune to pregnancy. Standing up during intercourse has no effect. A lot has been written about the timing of intercourse in order to choose the sex of the baby. Although some authors have claimed to show trends in large surveys, no indisputable advice can be given to the individual couple. Descriptions of sexing a fetus and diagnosing haemophilia in the womb will be found on p. 313 (see also Fig. 15.5).

Comparison of the methods

The contraceptive pill remains the most effective means of family planning, the chances of having a baby being minimal in regular users.

Too much fuss has been made of possible side-effects; the pill is now remarkably safe in women under 35 without a previous positive history (p. 307). All the other methods described, including the IUCD, carry a greater risk of pregnancy than the pill and many will be unacceptable to individuals. The rhythm or natural family planning method when it is followed extremely carefully in women with regular menstruation is reasonably effective. The least effective methods are probably douching, the use of spermicidal products on their own, and withdrawal.

It has only been possible here to outline the methods available and give some indications of their effectiveness and safety. Couples should discuss their needs with their family doctor or in a special clinic. In Great Britain clinics may be found in some hospitals, or the Family Planning Association may be approached directly. Wherever advice is sought consultations will be in private and discussion confidential.

Pregnancy

The bleeding disorders have no adverse effect on pregnancy. One of the natural results of being pregnant is that the levels of clotting factors rise progressively as the pregnancy advances. This means that even those carriers with unusually low factor levels at the start of their pregnancy will be safe by the time of delivery. There is no increased risk to the baby during a normal vaginal delivery, but of course the doctor in charge should know of the disorder so that the necessary treatment may be given in the event of a difficult birth or bleeding. Because of this mothers should be delivered in hospital and not at home. In general, babies who may have low levels of clotting factor (usually boys with haemophilia A or B) are better delivered by Caesarian section if a normal delivery is not possible. This is because a difficult forceps delivery may provoke bleeding into the head. Low forceps deliveries, in which the baby's head is protected as he is lifted out of the birth canal of a tired mother unable to complete delivery by herself, are safe. Rarely carriers are put at risk by bleeding after the third stage of labour, the delivery of the afterbirth. If this happens drugs will be given to help the womb contract and transfusion of the relevant factor may be required. As in many of the problems associated with bleeding disorders, to be forewarned is to be forearmed.

Artificial insemination
Artificial insemination by donor (AID)

This technique was developed primarily to help couples when the husband is infertile. Sperm from a carefully selected donor unknown to the couple are used to fertilize the mother's ovum. The procedure itself is relatively simple and some centres advocate its use at home rather than in a clinic. Obviously the husband is not the biological father of a resultant baby, whether male or female, and the most careful counselling is needed before a decision for AID is made.

The technique of artificial insemination becomes controversial when the woman chosen to carry the baby through pregnancy is not going to be the woman responsible for the child. Such a situation arises in the case of surrogate, or substitute, parenthood. A couple may decide to ask a surrogate mother to bear a child from AID using the husband's sperm. When the baby is born it is handed over to the couple and brought up as their own child. Although this practice is by no means new (nature took its course without the intervention of science long before AID) the subject is, not surprisingly, an emotional, legal, and sometimes a financial minefield. Once again, couples contemplating this course are well advised to seek the advice of a doctor with whom all aspects of their decision can be discussed.

AID (which should not be confused with the acquired immuno-deficiency syndrome, or AIDS, which is discussed in Chapter 12) may be relevant to families with haemophilia in three ways. Firstly, a severely affected man with haemophilia and his wife may decide that, in order to avoid the birth of a girl who must be a carrier of haemophilia, AID is the appropriate answer. Secondly, a woman known to be a carrier of severe haemophilia and her husband may wish to ask another woman to bear a child after artificial insemination by the husband's sperm. Thirdly, when a man with haemophilia is infected with HIV and the couple want a baby, AID may be the best answer because the wife will not be exposed to the risk of infection through unprotected intercourse with her husband.

In vitro fertilization

The term *in vitro* fertilization (IVF) means fertilization that takes place outside the body. The media have attached the label 'test tube babies' to the children born as a result of IVF; this suggests the Orwellian and frightening idea of batches of embryos growing to full baby size in laboratories. What really happens is that after a complex and beautifully timed work-up, an egg released from the mother's ovary is collected through a delicate suction apparatus. The egg is kept warm and healthy on a special medium and fertilized by the husband's sperm. After only a few divisions it is reimplanted into the mother's womb which has been prepared by hormone therapy to welcome it. This is embryo replacement, or ER. The pregnancy then proceeds in the normal way.

IVF techniques, pioneered by the British workers Steptoe and Edwards, have allowed previously infertile couples to bear their own children. If the only thing that is preventing pregnancy is blocked tubes which stop ova travelling from ovaries to womb, IVF might provide the answer.

The most recent example of the use of IVF is in families with a history of a sex-linked disorder like haemophilia. The very early embryo can now be sexed by removing only one cell. This removal has no detrimental effect on a resultant child because it is done before any cell differentiation has taken place. Knowing that a male embryo may have haemophilia, prospective parents (one of whom is a carrier), can opt for the implantation of a female embryo.

The picture becomes more complex when the sperms or the eggs used come from individuals other than the people who want to be parents. Here the procedure available is embryo transfer, or ET, an abbreviation which conjured up even more bizarre ideas following Steven Spielberg's film! In ET an egg is collected from one woman, fertilized, and then reimplanted in another woman. In fact a woman donates an egg for ET in exactly the same sense as a man donates sperm for AID. In terms of severe haemophilia, a couple, knowing that the wife is a carrier and not wanting to contemplate either having a haemophilic child or a termination, might decide to accept an egg from a donor, have it fertilized by the husband's sperm and then transferred to his wife's womb to develop and grow. In this case the wife is not the biological mother of the child, whether male or female.

Summary

Therefore, with these techniques, the possibilities for prospective parents in some countries are as follows:

1. Man with haemophilia + wife
   Fact: Each daughter must be a carrier.
   Possible course of action:
   (a) *In vitro* fertilization, sexing, and implantation of male embryos only.
   (b) Artificial insemination of wife with donor sperm.

2. Husband + carrier wife
   Fact: Each son has a 50:50 chance of inheriting haemophilia. Each daughter has a 50:50 chance of being a carrier.
   Possible courses of action:
   (a) *In vitro* fertilization, sexing, and implantation of female embryos only.

(b) Artificial insemination of surrogate mother who bears the child.

(c) In *vitro* fertilization of egg donated by another woman, which is implanted in wife's womb. She then bears the child.

(d) In the near future, it should be possible to diagnose haemophilia in the fertilized egg before implantation. This will allow couples to have children of either sex by embryo replacement using their own eggs and sperm.

The ability to perform IVF and ET has opened up a Pandora's box of possibilities, and some of these are very worrying. It is clear that legal guidelines for the procedures are needed, especially to cover experimental and possible modifications of the techniques and in the consequences of deep freezing and storage of embryos.

Diagnosis before birth

It is possible to diagnose severe haemophilia A, severe haemophilia B (Christmas disease), and severe von Willebrand's disease early enough in pregnancy to allow a couple the option of terminating an affected fetus. This can either be done by chorionic villus biopsy, or by fetoscopy (Fig. 15.5).

In the case of the haemophilias, if the family want another child and it is certain that the mother is a carrier, they will know that there is a 50:50 chance that any boy will be affected. They might decide that this risk is too great to take but would like a little girl, even though she will have a 50:50 chance of being a carrier herself.

In the case of the haemophilic father, it is known for certain that all his daughters must be carriers. A decision to have only sons, who will be unaffected providing his wife is not a carrier, will prevent the transmission of haemophilia to his grandchildren and thus to future generations.

In the case of severe dominant von Willebrand disease, if either parent has the disorder there is a 50:50 chance of a child of either sex inheriting the disorder. When von Willebrand disease is recessive, the chances of inheritance for a child of either sex are one in four.

The fetus can be sexed very early in pregnancy using the technique of chorionic villus sampling (placental biopsy) and, in an increasing number of cases, a specific diagnosis of haemophilia made by examination of the DNA. The technique is usually performed around the 8th–10th week but can be used later in pregnancy. Under careful ultrasonic scan control a thin tube is inserted via the vagina and cervical canal, or through the abdominal wall, into the womb. The tip of the tube enters the attachment of the placenta to the womb, and a tiny fragment of tissue is sucked into it; this tissue has the same genetic structure as the fetus. The risk to the fetus in this technique is around 3 per cent.

When chorionic villus sampling is either not available or not applicable because of the stage of pregnancy, safe and accurate sexing of the fetus is possible using high-definition ultrasound. This may show the anatomy of the fetus in sufficient detail to distinguish male from female genitalia. When this is not possible a painless procedure called amniocentesis can be carried out in hospital. At about the 15th week of pregnancy a thin needle is introduced through the mother's abdominal wall and into fluid surrounding the fetus in the womb. A small amount of fluid is withdrawn and examined. The fluid will contain some of the fetal cells and with special techniques these cells can be sexed.

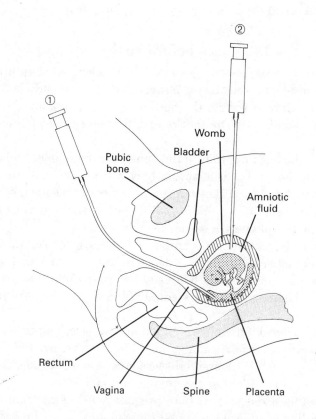

Fig. 15.5. A composite picture of two ways of studying the fetus in the womb. (1) shows chorionic villus biopsy, which can allow fetal sexing and haemophilia diagnosis early in pregnancy. The syringe is used to obtain tissue from the placenta which contains the same cells as the fetus. (2) Shows amniocentesis, which is used to obtain cells present in the fluid surrounding the fetus. Fetoscopy, during which a tiny specimen of fetal blood is taken from the umbilical cord, is a similar procedure.

The procedure is very unlikely to harm the mother or the fetus and pregnancy can continue in the normal way. In skilled hands the risk of amniocentesis to the fetus is around 1 per cent.

If there are legal grounds for terminating the pregnancy of a carrier (female) fetus, only fetal sexing is necessary. When the fetus is male and may have severe haemophilia A or B, or when a fetus of either sex may have severe von Willebrand disease, and DNA analysis from a chorionic villus sample has not been possible, fetoscopy is indicated. Fetoscopy, like amniocentesis, involves the introduction of a needle into the womb. However, this time the needle contains a bundle of glass fibres down which a light is shone, an optical system, and a tube through which a blood sample can be drawn. Fetoscopy is possible around the 18th to the 22nd week of pregnancy. It is performed under local anaesthetic and the instrument is guided into position using ultrasound, which shows a picture of the womb, fetus, and placenta on a television screen. Under direct vision, the point of the internal needle is introduced into an umbilical vein and a tiny specimen of fetal blood withdrawn for laboratory analysis. If the fetus is severely affected, termination is performed. Fetoscopy carries a 2 per cent risk of injury to the fetus.

Only the parents can decide whether or not to consider abortion, and they should accept no pressure from their relatives or neighbours. They must make up their own minds on the evidence available, and their decision must be private and respected.

Obviously there is no point in going ahead with any of the techniques described if termination is not contemplated, but a couple may change their mind between the procedures. If they do they should never be afraid to tell their doctor.

It should be remembered by those who oppose abortion for whatever reason that these techniques open the way for couples who would not otherwise have children because of the risk of haemophilia to decide on pregnancy in the knowledge that their child will not be affected.

Termination of pregnancy

Therapeutic abortion is now available in many countries for people with a medical or social reason. The means of recommendation for termination vary from country to country, but in general the first person to see is either the family doctor or the haemophilia centre specialist. The first step should be taken as early in pregnancy as possible to allow time for the necessary arrangements to be made. After the fourth month vaginal abortion becomes difficult and an operation may be necessary.

The actual procedure for abortion varies according to the gynaecologist in charge. The safest procedure for the patient will be chosen for that stage of pregnancy. All methods in early pregnancy are now relatively safe and quick, provided they are performed by a skilled practitioner in a recognized institution. The discovery of mifepristone (p. 308) will make things easy for many women because it puts the decision directly in their hands, allowing them to act independently of medical or other advice.

Sterilization

The most effective way to prevent pregnancy is to be sterilized. Operations for sterilization in either the man or the woman are simple, straightforward, and free from side-effects.

In the woman, the most usual procedure is to cut and tie off the tubes between the ovaries and the womb, thus blocking the passage of the eggs. There are different ways to do this, most through small cuts in the abdominal wall and some through the vagina. All methods are straightforward and quick, and usually require only a short time in hospital. There are no harmful after-effects and tubal tie does not result in an earlier menopause, in a lessening of femininity, or in any difference to the genitals or intercourse.

In the man the operation is called vasectomy, and involves the cutting and tying of the tubes (vas) which carry sperm from the testicles. Two tiny cuts are made in the scrotum to do this. The operation can be performed as an out-patient procedure and is virtually painless. Once again there are no harmful after-effects. The genitalia remain the same as before and erection and sexual function is unimpaired. As most of the ejaculate is made up of secretions from glands above the level of the vasectomy, there is no discernible difference in the amount of semen produced. The testicles continue to function normally, sperm being simply reabsorbed at their place of manufacture. The secretion of male hormone does not alter and there is therefore no change in 'manliness'.

Both in men and women sterilization may be followed by an increased enjoyment of sex, as the fear of pregnancy has been finally removed. Before agreeing to either form of sterilization the couple should be in no doubt that the procedure is final and will render the operated partner incapable of further parenthood. They will both have to sign a consent form to this effect.

Well-woman care

There has been a welcome shift of emphasis to preventive medicine in the past decade, and this is reflected in the concept of well-woman clinics. Everyone knows of the dangers of smoking and of excessive alcohol intake (especially during pregnancy), but there is still sometimes a reluctance to go for breast or cervical screening. Cancers of the breast and the cervix are curable, and the chances of cure are much better the earlier changes are found.

Breast examination

All women over the age of 20 years should examine their breasts once a month; some authorities suggest a 3-yearly medical check as well, especially for those who have used the contraceptive pill from an early age. Leaflets on breast self-examination are available from family doctors.

Cervical examination

All women over the age of 35 years should have cervical smear tests performed every 5 years. The timing of smears before 35 will depend on a particular woman in consultation with her family doctor or gynaecologist. It is probably best for a first smear to be done at the age of 20–22 years, each time a women attends for antenatal booking, and at least twice after the menopause. The procedure itself is painless and quick. The smear is taken directly from the cervix and, after special staining, is examined under a microscope by a specialist who can spot the changes that might herald cancer. A simple operation at this stage is curative.

Well-man care

As usual men are slow in catching up on their more health conscious female partners. Three aspects of male health should be checked regularly. They are:

- Blood fat and cholesterol levels, especially if there is a family history of early heart attack or stroke.
- Blood pressure, especially if there is a family history of hypertension.
- Testicular examination. Cancer of the testicle can be cured if caught early enough. If regular examination reveals a new lump in either testicle have a check-up from your doctor.

16

Towards a career: education and employment

The vital importance of providing the child with haemophilia or a related disorder with a good education cannot be overstressed. His or her future happiness and prosperity may depend on the choice of a satisfying career in keeping with any limitations imposed by the bleeding disorder. This often means brain rather than brawn, and jobs needing brain mean open competition with unaffected children and the passing of examinations.

Effects of haemophilia on schooling

Nowadays children with even severe haemophilia rarely miss school because of their disorder. In the past, before adequate treatment and counselling were available, about a third of the more severely affected haemophilic children missed 25 per cent of their schooling because of their bleeding. Many older people with haemophilia look back on childhood years of missed opportunities which resulted in poor quality jobs and long periods of unemployment.

The reasons for this are not difficult to find. Effective treatment with clotting factor concentrates has only been widely available within the past 20 years. Beforehand children used to be put to bed for weeks of rest after bleeds. Parents were reluctant to allow their affected children into the hurly-burly of normal schools, often overcrowded with big classes and poor supervision. In these conditions teachers had difficulty in maintaining a reasonable standard for the children without haemophilia. Those with any sort of impairment which prevented full attendance had little hope of keeping up with their peers. What schooling there was took place at home or in special schools for the physically impaired.

Nowadays the outlook for even the most severely affected child coming up to school age has never been brighter. The boy who is intellectually capable of going on to higher education should have as much chance as an unaffected child of doing so. And those of our patients who have gone on to medical school or other university faculties prove it!

What sort of school?

The easy answer to this question is 'the best possible'. Obviously the initial choice will depend on age, intelligence, locality, and perhaps religion, and in some cases family finance and tradition, just as it does for any child. In the case of the boy with haemophilia it may sometimes depend on the severity of his disorder as well.

The broad alternatives are either a normal school, a special school or, as a last resort, home tuition.

From the point of view of education there is no doubt that the normal school is much superior to the special school, and infinitely superior to home tuition, however dedicated the peripatetic teacher. From the point of view of the bleeding disorder, the great majority of affected children will do best at normal school. Very few will require more sheltered schooling, usually because of problems other than the haemophilia itself. Home tuition provides no long-term answer but can, very occasionally, be a useful short-term arrangement, particularly if carried out in close co-operation with the child's usual teachers.

Special schools

There are three main types of special school. The first caters for children with well-defined impairments, for instance the blind or the deaf. The second takes children with learning difficulties. The third type, which concerns us here, attempts to look after the needs of children with varied physical impairments. Although these children are supposed to be able to see, hear, and speak normally, and are meant to be of normal intelligence, it is rare to find a school for the physically impaired which fulfils all these conditions.

Because severe hereditary bleeding disorders are rare, schools only for haemophilic children are also rare. Some countries, notably France, have such schools, the authorities arguing that by combining medical and educational facilities, affected children have the best of both worlds. Elsewhere the haemophilic child will share his special schooling with children with many different physical impairments.

In the United Kingdom these schools may cater for either day pupils or boarders. Few local authority areas contain a special school suitable for haemophilic children, and families may have to travel long distances to be with their child. One boarding school in the United Kingdom, the Lord Mayor Treloar College, near Alton in Hampshire, has its own haemophilia centre, and another, Welburn Hall, at Kirbymoorside in Yorkshire, works in partnership with the Newcastle centre. If parents are advised that their child requires special schooling they should insist on talking to a doctor with experience in the management of haemophilia before a final decision on a particular school is reached.

Before starting school

Haemophilia is diagnosed in most affected children well before the start of their formal education. The severity of the disorder, both in terms of factor level and in terms of how it affects the boy, should be known. In the United Kingdom parents should make sure that they have been issued with a UK Health Department's Special Medical Card (Haemorrhagic States), and in other countries with either an appropriate card specific to that country's requirements or a World Federation of Hemophilia card. Cards are designed to include the diagnosis and factor level, the child's blood group, and details of inhibitors and allergies, together with where information and expert advice about him or her can be found. If a card has not been issued, parents should apply to their nearest haemophilia centre. The child should wear an identity disc.

This procedure is very important because a diagnosis of simply 'haemophilia' gives no clue to the severity of the disorder, and a boy with a mild factor deficiency may be treated by an Education Authority as though he was severely affected with tragic results.

In the year before their son is due to start school his parents should arrange to see the head teacher of the school he is expected to attend. They should tell the teacher the diagnosis, the effects of the disorder on the boy, and what precautions they take at home. The head teacher will work closely with the school doctor who, in the United Kingdom, is a representative of the community health services. It is at this stage that the first difficulty may arise.

It has been said many times already in this book that the haemophilias are rare. Advances in treatment are usually several years ahead of public knowledge and, indeed, these advances are sometimes so rapid that even medical textbooks are already out of date when they are published. Because of this, and because most doctors will not have met someone with a haemophilia since their student days, the school doctor may not be able to give the teachers up-to-date information about the disorder.

In addition, the whole subject has now been so coloured by the advent of AIDS that it is advisable that haemophilia centre staff are always involved at an early stage.

In Newcastle it is the practice of the centre staff to approach both the school teachers and the school doctor before the first term starts.

With the parents' consent, full details of their child's medical history, together with a description of haemophilia and its effects, are sent *via them* to the school. The parents approve the contents of the letter before it is seen by the school authorities. The teachers are invited to ask questions and, if necessary, a doctor, nurse, or social worker from the centre visits the school to talk over any problems with them.

The response to this combined approach from both parents and specialists is always extremely good. Everyone concerned with the child is fully in the picture and the teachers and school doctor become a welcome part of the team. The procedure is repeated each time the child changes school. The valuable publication of the UK Haemophilia Society, *Teaching boys with haemophilia*, should also be sent with the personal letter.

An example of the kind of information sent to the teacher and school doctor is shown below.

The Headmaster,
St. Matthew's School,
Newtown.

Dear Mr Smith,

Mark Williams, date of birth 19th January 1986
127 Anystreet, Newtown (telephone Newtown 856058)

I thought I should drop you a note about Tom. He is a fine boy who is developing normally. He has a younger sister, Anne, and the family live in Newtown, where his father is a plumber.

You will know from his parents that Tom has severe haemophilia. This is caused by the deficiency of one of the blood clotting ingredients. Because of this deficiency blood does not clot as quickly as it should following injury. Children with haemophilia bleed no faster than other children; their bleeds simply last longer. This distinction is an important one. It means that there is always plenty of time to seek help if it is needed, and it means that the first aid measures used with any child who is injured still apply.

Most bleeds in haemophilia are internal. If they show at all it is usually as bruises, most of which are superficial and of no consequence. Bleeds into muscles and joints do need treatment. Tom knows when he is bleeding and will tell his teacher.

Treatment of haemophilia itself is easy. The missing ingredient is replaced in the form of a blood product given into a vein. Tom's parents have been taught to do this themselves. If they are not available when Tom needs them, all you have to do is to ring the Centre; the names and telephone numbers of staff are appended. A member of staff is always available during school hours.

Tom is a healthy and active little boy. His parents have been very sensible, and have not tried to restrict his activities. We advise that he should be allowed to participate in everything his friends do at school. As he grows, he should be encouraged to develop skills in sport (especially swimming and team sports) and physical education, as well as crafts like woodwork and metalwork. If trips away from school are planned, the address of the nearest haemophilia centre and the name of the doctor concerned are available. For long trips or visits to other countries, we will supply a holiday kit, together with a letter for the Customs, details of treatment facilities and, when required, help with travel insurance.

The only restrictions affecting Tom are that he should not box or play rugby football. This is because of the special risks to the head and neck involved in these sports. Head injury in haemophilia can be serious and immediate referral to the centre is necessary.

Finally, it should be noted that haemophilia is not contagious; other children are not at risk!

We would be very grateful if you could keep us informed about Tom's progress, especially when careers guidance becomes necessary. We will be following him up regularly with his parents and are always pleased to answer any questions you or your teachers might have.

Yours sincerely,

Consultant, Haemophilia Centre.

Attached to this letter is a card with the relevant names and contact numbers on it. Ours looks like this:

NEWCASTLE HAEMOPHILIA REFERENCE CENTRE
ROYAL VICTORIA INFIRMARY, NEWCASTLE UPON TYNE, NE1 4LP.
Tel: 091 232 5131 Ext. 24170 Fax: 091 230 0651

Dr Jones and his Secretary may be contacted on extension 24170 on weekdays between the hours of 8.00 am and 4.30 pm.

Sister Fearns carries a bleep and may be contacted via the switchboard at the RVI, asking for Sister Fearns directly. Mrs Buzzard the Centre Physiotherapist is also available by bleep. The Haemophilia Centre is open for out-patient treatment on weekdays between the hours of 8.00 am and 4.30 pm.

Our Social Workers, Mrs Latimer and Miss Fraser may be contacted in the Centre on extension 24481 or 25181 on weekdays between the hours of 8.00 am and 5.00 pm.

At night or weekends patients should contact Ward 16 North (Extension 25016 children) or John Hall Ward (Extension 25013 adults).

In addition, the general background information shown on pp. 33−6 may be handy for the boy's teachers.

Reasons for special schooling

The haemophilic child going to a normal school is growing up in the community with his friends. His education is likely to be both better and more broadly based than that of the child in a special school. Why then recommend special schooling for some children? There are five main reasons:

- disruption of normal school by very frequent bleeds;
- the presence of high-titre inhibitors, making conventional treatment difficult;
- another medical problem in addition to haemophilia;
- problems within the child's family;
- distance between home and hospital.

With the increase in the number of families using home therapy, the frequency of bleeding and the problem of travelling long distances for treatment have gradually disappeared. Some children with inhibitors who bleed frequently and get well only slowly are possibly better off in a residential school where teaching can continue during recovery periods. Special schooling for children with multiple disabilities will depend on the extent and severity of their impairment.

The family reasons for special schooling include parental illness, bereavement, separation, and divorce. They also include the occasional family who, for very good and honest reasons, have fallen into the trap of overprotecting the child.

The pattern here is familiar to those who work with chronic illness in childhood. Following the diagnosis of the rare disorder the family receives little or no help with the process of living. In the case of haemophilia they may be able to get treatment for acute bleeds, but find themselves on their own when it comes to advice about the many facets which go to make up the world of a developing haemophilic child. They may even receive thoroughly bad advice or actual rejection from the professional people they approach for help. This is because the haemophilias are rare, and most doctors and social workers will not have met an affected person.

After months or years of trying, the inevitable happens. The family close round the child, often treating his muscle and joint bleeds themselves by putting him to bed when they occur. To prevent these painful episodes the child is restricted to playing with carefully selected neighbours, or even confined to the house. Home tuition takes the place of proper schooling. Realizing that the restrictions are unfair to a growing child compensations are introduced. Spoiling with expensive toys and special foods begins. Allowances are made for poor scholastic progress or bad behaviour 'because of his bleeding'. Gradually the child's world, which should be expanding, dwindles. Superficially everyone is happy and contented—the parents because they are making sacrifices in terms of time, money, and affection, and the child because nothing is demanded from him in return.

The outcome of all this is an adult with no qualifications who is totally dependent on his relatives and the State. He may or may not be physically crippled, but he is certainly crippled in mental, spiritual, and social terms.

There is usually only one answer to the problem, and it may involve a certain amount of 'being cruel to be kind'. The vicious circle has to be broken, and the sooner this happens in a child's life the better. Residential schooling sometimes provides the answer by giving the child a change of scene, and the parents respite, and time to start a normal life again. Most mothers and fathers realize the dangers and are pleased to have the opportunity to really help their child find his feet in the world. The results in terms of both scholastic achievement and the child's future personality are excellent. It is in these families especially that breadth of education, rather than the cramming of heads full of facts and figures, is truly important.

Employment

Well before he leaves school arrangements should have been made for the future career of the youngster with haemophilia. The end of compulsory education is one of the danger points in his life. It is known that the future prospects of any young adult who leaves school without the prospect of a proper job, and fails to find employment within a few months, are poor. With the passing of time the chances of his drifting through a succession of low quality, poorly paid jobs increase. The man with haemophilia has the additional burden of his disorder to contend with. For the first time in his life he may be in direct competition with unaffected men. All other things being equal, an employer is more likely to appoint a man without potential disability than the man with haemophilia. It is not surprising that at least 16 per cent of those with haemophilia hide the diagnosis from their employer.

Choice of career

The choice of career will depend upon the severity of the bleeding disorder. Mildly affected people, who only run into trouble after severe injury or after surgery, need have no restrictions. Some people with haemophilia, although 'severe' by laboratory standards (under 1 per cent factor VIII level) never have any trouble with bleeding, and regard themselves as normal from the point of view of work, and the same goes for most people with von Willebrand disease.

Certain occupations must be excluded from the career list of anyone with a proven diagnosis of severe forms of haemophilia A or B, von Willebrand disease, or another factor or platelet disorder which results in excessive bruising or prolonged bleeding. These occupations include the armed forces and the emergency services and, in some countries, the civil or diplomatic service. The reasoning behind these restrictions is obvious. In emergency, especially when far from home, it may be very difficult to get treatment to someone with haemophilia who needs it. In trying to do so the welfare of others may also be put at risk.

The second group of jobs which should obviously be avoided are those which will inevitably result in bleeding. In the past, men with haemophilia have been forced to take some of these occupations for want of any other work which does not require a qualification. They include mining, heavy labouring, and other work involving prolonged physical effort in awkward situations, for instance on trawlers.

In view of these restrictions, people with haemophilia sometimes ask whether, in the context of employment, they should regard themselves as 'disabled persons'. In the United Kingdom any person who, on account of injury, disease, or congenital deformity is substantially handicapped in obtaining or keeping employment may apply for inclusion in a Register of Disabled Persons. Application is made to a Disability Employment Adviser at any Job Centre. Unfortunately this procedure is no guarantee of a job; the employment of disabled people and able-bodied people is governed by the economic rules of supply and demand. Although attempts have been made in law to include a minimum number of disabled people in every firm of a certain size, there are too many difficulties and loopholes to make such a scheme effective.

Throughout this book emphasis has been placed on *normality*—the haemophilic child should be brought up in the same way as a normal boy in a normal environment. The same is true with regard to employment. The labels impaired, disabled, or handicapped confer no special status or automatic benefit. They should be reserved for use when all else fails, when they might serve as weapons to force an issue.

Nowadays, severely affected people should leave school with little if any impairment. They may have some restriction of movement in the elbow and ankle joints and possibly arthritis in one knee, which will respond to surgery later. Chronic pain is very rare at this age, and the only things likely to interfere with work are acute bleeds. With effective home therapy and prophylaxis these should present few problems.

The ideal job for someone with severe haemophilia

In an ideal situation the following conditions of employment would apply:

- He would enjoy work.
- The employer would know of the haemophilia and have accurate and up-to-date information about the employee's disorder.
- There would be an arrangement to provide cover during unavoidable absence.
- Facilities for efficient and speedy treatment should be near at hand.
- The job should not involve work likely to put heavy strain on joints and musculature, and should be in a warm, dry environment if the employee has chronic arthritis.
- The job should allow movement and not confine the employee to long periods of sitting or standing in one position.
- The rates of pay and conditions of service should be the same as those for non-affected employees.

Career guidance

Guidance towards a child's future career begins at school in the early teens. The teachers will be becoming aware of his special gifts and limitations, and start to discuss these with him. From now on the wide choice is gradually condensed to a number of options. These will depend on his intelligence and aptitude, and the type of school he attends. Parents should be gently assessing the future of their haemophilic child as well, and school Open Days or other invitations to meet with the staff give them an opportunity to talk things over with the teachers. Work experience days help youngsters choose their eventual career.

The statutory school leaving age in the United Kingdom is 16 years. In the previous year each child should be interviewed by a Careers Officer from the local authority. He, or a careers adviser or teacher from the school, will speak to the parents and the child together. If this does not happen parents should ask what advice is being given to their child.

In referring to the list of requirements for a job suited to the severely affected boy, it will be apparent that the nearest occupations to the 'ideal' are those requiring special qualifications. In Britain the first of these qualifications would be the General Certificate of Secondary Education (GCSE).

Although these examinations are normally taken before the statutory school leaving age there is no age bar. If a child's education has been affected by his haemophilia his parents should very seriously consider the possibility of his education being extended for as long as is required for him to pass the examinations. Where this further period of education will be spent will depend on the school and the provisions of the local education authority. Pupils at special schools may be transferred to normal schools in their own neighbourhoods. Those at comprehensive schools may be able to stay on at the head teacher's discretion, or, if they are bright enough, obtain a transfer to a school or college with sixth-form facilities.

Only when it is absolutely necessary should any disabled or potentially disabled person leave his formal education at the statutory school-leaving age. He may sit his examinations later than usual, or resit if he has already failed some subjects. When he is not likely to make the grade in terms of examination success, an extended education gives him time to mature through experience with projects like the Duke of Edinburgh's Award Scheme, youth club movements, and work experience outside the classroom. He will be working and living with unaffected boys, and his experiences will increase his independence, and be invaluable to him in later life.

Job placement

Even in these days of high unemployment, no haemophilic child should leave school without every attempt at obtaining employment. The careers of the brightest children will have been mapped out after GCSE, and they will go on to qualify for higher education. The others will, as has already been indicated, receive guidance from a Careers Officer. Parents should ensure that these wheels are turning well in advance of school leaving. If necessary, they should ask about vocational guidance if arrangements for this have not already been undertaken.

Here the haemophilia centre staff may be able to help. The major hospitals have access to educational psychology facilities either directly or through the social services. Parents sometimes confuse 'psychology' with 'psychiatry' and worry about the erroneous implication that something is mentally wrong with a child referred for psychological assessment. Nothing could be further from the truth. Psychology is the science of measuring intelligence and personality, and the educational psychologist is experienced in matching these findings to suitable careers. When his service includes job placement as well as advice, assessment is particularly valuable. He may also be able to give guidance on interview techniques.

The older person with haemophilia

There is now a wealth of facilities for further education in Britain. Evening classes, sandwich courses, and the Open University provide everyone capable of persistence with the opportunity to further their education and obtain qualifications they may have missed in childhood. In many cases Government grants are available, and some jobs provide for further education in the course of normal employment. No one is too old to learn.

For others the Disablement Employment Adviser may provide an answer. He is able to refer people for rehabilitation or vocational training in specific jobs, which will vary from area to area. Disablement Employment Advisers are members of Placing, Assessment, and Counselling Teams (PACT) and can be contacted via local Job Centres or Unemployment Benefit offices. Once again the staff of the haemophilia centre or the Social Services may be able to arrange vocational assessment for the man who is unsure of the best way to proceed, and the Haemophilia Society is always ready to give individual advice.

Specific problems

Should an employer know? Once someone with haemophilia has trudged from interview to interview without success, one of his reactions will be to hide his disorder from his potential employers in order to compete on the same level as other applicants for the job. Some people manage to hide their disability successfully for years, but if the job is demanding and the disorder severe the deception can become a very frustrating one. It often pays to ask the staff of the haemophilia centre, or the doctor who knows the man well, to talk in confidence to a potential employer, and to send him details about the up-to-date management of haemophilia. Employers who know only that haemophilia has something to do with bleeding are unlikely to be keen to give a job to someone with haemophilia. Once they know what haemophilia really means they are usually sympathetic and often go out of their way to be helpful.

What about treatment for acute bleeds at work? Few people have not been trained to inject themselves and home therapy is just as possible at work. If for some reason home therapy is not practised, arrangements can be made for treatment at the nearest hospital. Large firms have casualty stations, staffed by trained nurses, and it may be possible for treatment facilities to be set up by them in conjunction with the haemophilia centre.

There follows the updated text of a booklet originally prepared for people with haemophilia in the United Kingdom in 1980. Some of the tips it contains may help those finding it difficult to get a job.

Haemophilia and work
Introduction

Most people with haemophilia find a job relatively easily, stay in it for many years, and are content with their work. Others find life more difficult. The difference is not necessarily to do with the severity of the condition; there's more to it than that.

It's mainly to do with knowing about haemophilia, recognizing your limitations, and learning to live with them. It's also to do with knowing how to explain these things to a prospective employer; he may appear unsympathetic, but that is probably because he doesn't really know what haemophilia means. Who's going to tell him if you don't?

This text has been written to help you choose the right job, to get it, and then to cope efficiently with your special problems at work. A booklet can only help—by giving you advice and pointing you in the right direction. The rest is up to you!

Explaining what haemophilia is

You may need to explain about haemophilia at an interview. Although you know a lot about your own condition, it's sometimes difficult to choose the right words for the layman. Here are some facts and tips which may help you.

- Haemophilia is a condition in which one of the clotting factors in the blood is lacking. This means that bleeding after an injury tends to go on slowly and persistently for many hours rather than stopping in a few minutes. The volume of blood is not greater, it just goes on for longer. It is important to get this point across, because most people think that someone with haemophilia will bleed to death from a pin-prick! It's also important to stress that people with haemophilia pose no health risk to others at work.

- It also seems to be necessary to point out that bleeding, when it does occur, is more often than not internal and into a joint or muscle. Usually only the affected person knows himself it's happening and it's not visible to the non-expert.

- Most people will be surprised to know that superficial cuts and grazes usually stop bleeding quite quickly after normal first aid, and so this is another important message to put over.

- Haemophilia is an inherited condition in most cases; it is life-long, and at present can't be cured. The symptoms can, however, be completely corrected by the injection of a concentrate of the missing clotting factor into a vein. The injection can be given to stop bleeding or to prevent it from starting in the first place.

- Haemophilia if of variable severity. A mildly affected person may never receive an injection of concentrate until he has to have a tooth out or undergo an operation—the injection is then given beforehand to prevent bleeding.

- A severely affected person may, however, require an injection several times a month. He is then usually trained to give his own injections. It might be an idea to make the comparison with a diabetic with insulin, since most people already know about that. Thus, even a severely affected man can lead a near normal life and offer as much as the next man to the working community.

- There are about 6000 males with haemophilia in the United Kingdom. Each man keeps in close touch with his local haemophilia centre, where expert help is immediately available round the clock.

What job to go for

Whether you're looking for your first job, or you're temporarily employed, or if you just feel you want a more interesting job, try to work out what you really want to do and go for that first. Start at the top. Don't be discouraged and take second best until you've explored all the possibilities. Nowadays there's lots of help to be had if you know where to look for it.

People who might help

The place to start is your own haemophilia centre. The social worker there will know your history and will therefore be able to give you local names and addresses and leaflets relevant to your particular needs. Contact all the people whose names you're given, no matter how unpromising they may seem; one contact leads to another and practical help often comes from the most unexpected sources.

Here are some short notes on the ways in which various people might be able to help you.

Disability Employment Adviser

Your local DEA can give you practical help in finding a suitable job in your area, and advice about the general suitability of a particular job you might have in mind.

They will also give you information about facilities and help available to you through the Department of Employment, (listed in the telephone directory under 'Employment, Department of').

Employment Medical Advisory Service

You can always get advice about the most suitable type of work for you on the grounds of your haemophilia from the Employment Medical or Nursing Adviser at your local branch of the Health and Safety Executive. Again, you can get their address and number from the telephone book.

Help for young people

If you're under 18 you can get advice about further education and training, as well as practical help in finding a job, from the Careers Officer at your Local Education Authority. You can go and see the Careers Officer even after leaving school if you're still having difficulties.

In most areas there are now also specialist Careers Officers who can give extra help and advice if you have additional problems because of your haemophilia.

The address of the nearest Careers Advisory Service is in the telephone directory under Education Department, which in turn can be found under the name of your town or district.

You can also get additional advice, if you're over 16, from the nearest Occupational Guidance Unit—the local Job Centre or Employment Office will give you the address.

The Youth Training Scheme offers both short (13 week) assessment courses for those unsure of what they want to do, and two-year training programmes for specific career choices. Why not go and find out more about them?

Any information or advice people have to offer, to help you find the right job, is always worth following up. Ring first and make an appointment whenever you can, so that the officer concerned can refer to his own information sheets on haemophilia and employment opportunities before he sees you.

How to help yourself

Dedication and enthusiasm on your part are necessary too. Scour the newspapers and keep up-to-date with what your Job Centre has to offer. Follow up every possibility. You should feel that being out of work gives you a full-time occupation—looking for a job or retraining yourself for a new career.

Some firms do not advertise vacancies. Instead they rely on self-referral by people seeking a job with them. So advertise yourself; write round the firms which you would like to work for. Your name may be kept on a list, and you may gain marks for initiative.

Don't be discouraged if you don't seem to be getting anywhere and never blame lack of success on your condition. The fault probably lies with the unfortunate state of the employment market.

Remember that the job market varies from week to week. If you're looking for a job, it's up to you to keep track of what's going on in the business world. Opportunities also vary from place to place. If you're in an area of high unemployment, are you prepared to move to where the jobs are? Haemophilia treatment is available nationwide.

Put as much effort and determination into job-hunting as you plan to put into the job itself—even if it takes you several months to get one.

Interviews

Go to all the interviews you can. They are tricky things for most people and you need to gain experience; there's no other way to get it.

At an interview you need to sell yourself. Your aim has to be to come over, not just as well as the other candidates, but better—otherwise there's no reason for a prospective employer to choose you.

If you suspect that you're not very good at selling yourself, or that perhaps you over-sell yourself, get a book on interview technique from the library, and ask your friends and relations about their experiences at interviews.

It's important not to feel discouraged if you get as far as the interview but don't get the job. You may have been pipped at the post, or you may have made a mess of it. If you have, you'll know, and you'll handle the next one better.

Practice makes perfect—therefore go to as many interviews as you can. Sooner or later you'll find yourself at the right one.

Telling your employer

The decision whether to tell your employer about your condition or not is yours and, whatever you decide to do, this will be respected by your haemophilia centre team.

There have been instances in the past of people with mild or moderate haemophilia who have stayed in the same job for years without anyone knowing. If you are only mildly affected, you might be able to get away with it too, *but it's not recommended.*

You can't be sure how much your employer already knows and, although you may keep your secret for a time, it might come out and put you in a very embarrassing position. You could even be sacked. It would then be very difficult to get another job, especially if a reference is required.

By law, every employer now has the responsibility at work not to endanger the health and safety of his employees, and this should be remembered. Help him, therefore, by telling him about your condition and any special requirements you may have. In this way you can ensure that any equipment you use for self-treatment is stored and disposed of properly. If you wish, someone from your haemophilia centre will go with you to talk over any problems with your employer or his personnel officer.

A great deal is being done to encourage firms to employ people with conditions like haemophilia. Companies are responding to these appeals, and so it should pay to come clean.

Being secretive about your haemophilia can make it impossible to work with your colleagues in a relaxed atmosphere.

Most employers and colleagues at work tend to be very helpful once they really understand what haemophilia is, and seeing how well you cope with it will probably win their admiration.

Finally, remember that, whilst it is wise to tell a future employer about haemophilia, there is no need to disclose your HIV status—the law is on your side.

Knowing your limitations

It is important to be objective—to be able to look at yourself and recognize your limitations. You can't work efficiently or comfortably—nobody can—if a job is too testing, either physically or mentally. Don't just think about what you might be able to do on a good day, consider what would be impossible on a bad one.

Your own experience, and the advice of your haemophilia centre staff, will tell you what you are physically capable of at present and what you will realistically be capable of in the future. Find out how much physical exertion is involved in a job and make sure that what you're going for is within your scope.

If you're honest with yourself about these things you will find that employers sympathize with your aims and ambitions and will try to help. Learning to live within one's limitations is something everyone has to do, and in your case it's particularly important. It's the starting point for doing a job well.

Try to assess the mental challenge of the job too. Working up to the limit is something many people find stimulating, but if you take on a job which is likely to be constantly stressful, be sure you can cope with it in addition to your haemophilia.

Implications at work

If you're mildly affected you will rarely require hospital treatment. An employer shouldn't take exception to occasional absences for treatment or to attend an out-patient clinic, especially when he realizes that this will ensure that you are as fit as possible and able to work to your full capacity.

If you're more severely affected, or live a long way from your haemophilia centre, you are likely to have been trained in self-treatment. Be sure that you explain carefully, at an interview and in the job, what this means.

First, it means that you have been trained to recognize a bleed as soon as it starts and, by injecting yourself with concentrate, to put things right and get back to work within a quarter of an hour. Second, it means that you need a lockable cupboard at work to store your treatment kit, a clean working surface, and a wash-basin. This is not a lot to ask.

Self-treatment at work

If you can treat yourself, then don't restrict yourself to self-treatment at home—that's doing only half the job. These practical points may be useful when you start to treat yourself at work.

- Don't feel shy about your injection and attempt to do it on the quiet, or in an unsuitable corner, or in a hurry. Be confident and proud of your achievement—few of your colleagues could cope with it.
- Your special requirements are minimal, and so ask for them and explain their purpose.
- The ideal place for your injection is a first-aid room, or an office, or some other clean area with a clear working surface and access to hot water. Ideally someone should be on hand—either the works nurse or a friend—in case of any allergic reaction.
- Keep your treatment kit in a locker or lockable cupboard and keep a spare key at work.
- If there is a domestic type refrigerator (4 °C) which is not used for food, then store the unopened bottles in the main part of the cabinet (not in the freezing compartment).
- It is permissible to store most concentrates for short periods at room temperature; read the directions leaflet that comes with your concentrate. You should not store more than three or four bottles at a time. Use your bottles in date-order to ensure rapid turnover. Avoid leaving bottles in direct sunlight or in a car unless they are protected by an insulated container. Ask your haemophilia centre director for advice about your particular concentrate if you are unsure.
- Always clear up very carefully afterwards. Put used needles in the special container provided and put the rest of the equipment in a plastic disposal bag. Seal this carefully and take it home for subsequent disposal, as advised by your haemophilia centre. Never leave needles or sharps lying around; they may injure someone else.
- Remember that you may well have become immune to infections like hepatitis, but other people have no such protection.
- Always maintain the highest possible standards in your injection technique. A dirty work surface, a hurried technique, or the

accidental touching of the clean needle, must never be accepted. Remember that you're in a working atmosphere, as opposed to in hospital or at home. So ensure that you get the best possible hygienic conditions.

Insurance, superannuation, and pensions

Most people with haemophilia who are HIV negative have no difficulty in joining an insurance, superannuation, or pension scheme run by their employers. If difficulties do arise, however, the insurance expert at the haemophilia society can often give sound and practical help.

Practical points to bear in mind

Getting a job is one thing—getting to it is another. Check beforehand with your haemophilia centre if you are eligible for travel assistance. If you do go to work by car, is there the possibility of a parking space near your department entrance?

If you're on self-treatment, make sure that there are adequate facilities and ask to see them.

It's always better to check upon these details before you accept or start a job.

. . . and a last word

It is hoped that this text will help you to find a job you really enjoy doing or make life easier in your present job. It was originally produced by a working party for the United Kingdom Haemophilia Centre Directors. If your own experiences have given you ideas for improvements to this booklet which may help other haemophilic job-seekers in the future, then please tell your haemophilia centre staff about them.

PART VI

Towards a cure

17

History, and the hope for gene therapy

In March 1992, whilst I was attending a World Federation of Hemophilia meeting in America, news came through that two youngsters with haemophilia B had received gene therapy in Shanghai. The news, although not confirmed, was exciting for two reasons. First, it suggested that the Chinese might have won the race to cure one form of haemophilia. Second, in highlighting the gap between developed and developing countries, it held out enormous promise for people previously without access to treatment. If the experiment was successful the Chinese would, in one leap, have cleared the gulf between diagnosis and cure without having to think in terms of conventional clotting factor therapy. The savings, both in terms of human health and happiness and in terms of economic resource, were obvious.

Whatever the final outcome of the Chinese initiative, the road towards cure for both haemophilia A and B seems clear. In October 1993 doctors in the USA reported the results of their first attempt to cure dogs with haemophilia B. Their results were encouraging, but suggested that there was some way still to go before sufficient clotting factor could be expressed from new genetic instructions implanted in humans with haemophilia. The book detailing the history of haemophilia cannot yet be closed. In the interim it is interesting to look back and see how far we have come, and to look at how gene therapy may eventually be achieved.

The history of haemophilia

Primitive man recognized the change from the flowing blood of injury to the sticky congealed material of a clot. The ancient Greeks certainly knew of, and wrote about, the change. They saw that as a clot gradually shrank it exuded the yellowish liquid we now call serum, and they observed the white dense fibrin left after a clot is washed in water. Many of their medical theories were based on observations of blood and bleeding. The father of medicine, Hippocrates, who lived between 460 and 370 BC, suggested that clotting might be due to the cooling of blood as it left the warmth of the body. Bleeding disorders were recognized very early in history; Jewish boys from families of bleeders were excused the rite of circumcision.

Although little more was known about clotting until the 19th century, two earlier series of experiments are of interest because they anticipated modern knowledge. The first, recorded by Samuel Pepys in his diary, took place on the 14th November 1666 at Gresham College in England, and provides us with one of the first accounts of blood transfusion. In an experiment one dog was transfused with the blood of another. Pepys remarked:

> This did give occasion to many pretty wishes, as of the blood
> of a Quaker into an Archbishop an' such like; but may if it
> takes be of mighty use to a man's health for the mending of
> bad blood by borrowing from a better body.

The second was in the latter part of the 18th century when animal experiments in Dr William Hunter's anatomy school showed that plasma, not red cells, was involved in clotting, that cooling delayed rather than led to clotting, and that the blood vessel walls were in some way responsible for keeping circulating blood in a fluid state.

Although most doctors in the 19th century were more interested in relieving patients of blood by lancing veins or applying leeches than in attempting to put it back or to understand clotting, a number of advances were made. An obstetrician, James Blundell, of Guy's Hospital in London, performed several successful blood transfusions. Later Dr Alfred Higginson of Liverpool reported further successes using his own invention of a valved syringe; his patients were women who had bled excessively at childbirth.

In the course of the century it became known that tissue fluid would initiate clotting in blood, that proteins in the plasma were involved in the process and that one of them, fibrinogen, was converted to fibrin by another protein, thrombin. Failure of blood to clot following the addition of chemicals which removed calcium was shown, the blood later clotting if calcium was put back. This was a crucial discovery as on it depend many of the laboratory tests for clotting in use today.

In 1905 a German, Paul Morawitz, published a paper on clotting which provided the basis from which modern theory and practice has evolved. Thus at the turn of the century it was known that the blood protein prothrombin was converted to thrombin by the action of calcium and a substance in the blood named thrombokinase. Thrombin then converted fibrinogen to fibrin.

These reactions are the end stages of coagulation recognized today. In the years since Morawitz's paper, thrombokinase has been shown to be composed of a number of different proteins which all act together in health to spark off the final appearance of a fibrin clot.

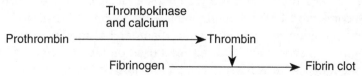

The discovery of the different proteins and how they work together was helped by the introduction of tests which enabled scientists to measure clotting reactions in the laboratory, by the isolation of anticoagulants, and by the study of patients with bleeding disorders. The first tests, still in use today, were the result of work by the American investigators A. J. Quick and Warner, Brinkhous, and Smith. They were soon employed in discovering why cattle in Alberta and North Dakota were dying from a bleeding disorder. The cause was traced to a substance in the sweet clover eaten by the cattle, and this substance, called the coumarin, was isolated by Link and his colleagues in Wisconsin. From it were derived the first anticoagulants that could be taken by mouth. One of these, warfarin, was initially used as a rat poison—it still is today—but soon became one of the most widely used drugs for people at risk from thrombosis. Its name recalls its birthplace: 'W' for Wisconsin, 'a' for alumini, 'r' for research, 'f' for foundation, and 'arin' for coumarin.

The coumarins were not the first anticoagulants of use to humans. The study of clotting by Howell in America led to a student, Jay McLean, trying to isolate one of the substances which started the process. This substance was known to be present in the tissues and McLean tried to find it in liver. Instead, in 1916, he discovered the anticoagulant heparin—'*hepar*' is Greek for 'liver'. Heparin acts quickly and is very important in a multitude of ways, including the treatment of patients who require heart surgery or kidney machine dialysis.

In the early experiments rabbit brain had been used to supply thrombokinase to the mixture under test in the laboratory. In fact other tissues would have done just as well, but this was not known when, in 1943, a patient in Oslo presented with a bleeding disorder. Her doctor risked his life by defying the laws of the Nazi Occupation in Norway, and cycled out into the countryside to catch rabbits. In a series of brilliant experiments Dr Owren proved that his patient had a previously unknown condition. She was not deficient in the four factors then recognized—prothrombin, fibrinogen, calcium, and 'tissue fluid'—the cause of her illness was therefore due to a fifth factor, later named factor V. We now know that Dr Owren risked his life unnecessarily. Thankfully he survived the Occupation and went on to extend our knowledge of clotting, and to introduce one of the ways of measuring the response of patients to anticoagulants, allowing them to be treated in safety.

The pattern of the clotting mechanism recognized at the time of the Second World War was obviously incomplete. Five factors were known but none of them fitted the disorder haemophilia, and none explained a discrepancy between tests used on blood which had been anticoagulated with warfarin. The factor causing the discrepancy was discovered in the 1940s and later named factor VII (factor VI was dropped from the list); the first patient known to have VII deficiency was a little girl described in 1951. Just before the war the fact that haemophilia was caused by a clotting factor deficiency present in plasma was recognized. The factor was called anti-haemophilic globulin (AHG), later to be known as factor VIII.

Christmas disease (Haemophilia B) was discovered by Rosemary Biggs and her associates in Oxford in 1952. Christmas was the name of the first patient in whom they proved the deficiency of a new factor (IX). This demonstration explained the previously thorny problem of why the clotting defect in the blood of some people with haemophilia was corrected by mixing it with the blood of others thought to have the same condition. In these experiments factor VIII deficient blood and factor IX deficient blood were being mixed and were correcting each other. Because VIII and IX deficiencies are inherited in the same way the difference had not been apparent in the family histories, and to this day we cannot be sure from which type of haemophilia the descendants of Queen Victoria suffered.

The 1950s saw the discovery of the other factors, and all of them were given their Roman numbers in 1961 when, in order to avoid the confusion of names given by different scientists in different countries, an International Committee introduced the more sensible nomenclature. There are now 12 numbered on the field of play and a number of others on the sidelines.

How they all interact and lead to the fibrin goal is still a matter of speculation and experiment. The important thing is that their recognition allowed scientists and doctors to prescribe the right treatment for each of the deficiencies which cause a bleeding disorder.

It is probable that in the near future the clotting factors will be renamed or numbered again. This move will be dictated both by increasing knowledge of clotting, and by the requirements of computer technology which does not like Roman numerals.

Haemophilia

The history of haemophilia is probably as old as the history of man, but the disorder would not have been so widely recognized if it had not been for the descendants of Queen Victoria of England. Her family tree is shown in Fig. 17.1.

Victoria was the granddaughter of George III. She was born in 1819, the only child of Edward, Duke of Kent, and Victoria, Princess of Saxe-Coburg. She succeeded her uncle, William IV, to the throne in 1837 and three years later married Albert, son of the Duke of Saxe-Coburg-Gotha. They had nine children, four sons and five daughters. Of the nine, one son, Leopold, had haemophilia, and at least two daughters, Alice and Beatrice were carriers.

Leopold was the youngest son. During his birth on 7th April 1853, chloroform was administered to Victoria by Dr John Snow of Edinburgh, a landmark in the history of early general anaesthesia. Leopold was clinically severely affected. He bruised easily and suffered recurrent haemarthroses which left him with a permanently affected knee. When he was 15 the Queen bestowed on him the Order of the Garter, wishing 'to give him this encouragement and pleasure, as he has so many privations and disappointments'.

At 26 Leopold was prevented from representing the Queen at the opening of the first Australian International Exposition. Writing to Prime Minister Disraeli, Victoria commented:

> She cannot bring herself to send a very delicate son, who has been four or five times at death's door, who is never hardly a few months without being laid up, to a great distance, to a climate to which he is a stranger and to expose him to dangers which he may not be able to avert. Even if he did not suffer, the terrible anxiety which the Queen would undergo would unfit her for her duties at home and might undermine her health.

The Queen was right. Leopold's place in Australia was taken by his brother, Alfred, who was shot. The bullet was removed and he recovered; the probe used in the operation still hangs on the wall in the hospital named after him in Sydney. If it had been Leopold there is little doubt that he would have bled to death, the victim of assassination.

Leopold married in 1882 at the age of 29 years. His bride, Helena of

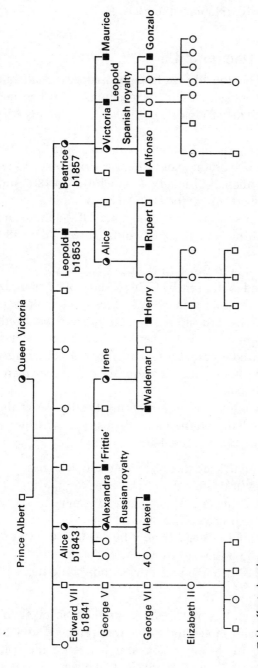

Fig. 17.1. The British Royal Family tree. Haemophilia in Queen Victoria's family. Edward VII did not have haemophilia, therefore the present British Royal Family cannot have inherited the disorder.

□ Unaffected male
■ Affected male
○ Normal female
◐ Carrier female

Waldbeck, bore him two children, a daughter, Alice, and a son. Before the son was born Leopold died after a fall in Cannes; he was 31. Victoria wrote: '. . . for dear Leopold himself we could not repine; there was such a restless longing for what he could not have that seemed to increase rather than lessen'. His daughter Alice was, of course, a carrier and at least one of her descendants, Rupert, was haemophilic. Rupert died after a car crash in 1928.

Queen Victoria's carrier daughters were Alice, born in 1843, and Beatrice, born in 1857. Princess Alice had seven children, of whom one, Frederick had haemophilia, and two, Alix and Irene, were carriers. Frederick, or 'Frittie', died when he was 3-years-old after a fall from a window. Alix became Czarina of Russia when she married Nicholas in 1894, taking the name Alexandra Fedorovna. One child, Alexei, had haemophilia, but we shall probably never know if any of his sisters were carriers. When the remains of the family were found and identified by genetic fingerprinting after the disintegration of the Soviet Union two bodies, including that of Alexis, were missing.

Princess Irene married her cousin Henry of Prussia and, of their three sons, two were affected. Waldemar died in 1945, aged 56. His brother Henry died of bleeding aged 4, after a short life hidden from society.

The second of Victoria's carrier daughters, Beatrice, passed the haemophilia gene to the Spanish royal family following her marriage to the Prince of Battenberg in 1885. The family name was changed to Mountbatten during the First World War at the instigation of George V. The Mountbattens had four children. Leopold and Maurice had haemophilia. Incredibly both served in the war, Maurice being killed in action at Ypres in 1914. Leopold lived to 33, when he died after an operation.

Their sister, Victoria, married Alfonso XIII of Spain, bearing him five sons and two daughters. The youngest son, Gonzalo, had haemophilia; he died after a car crash in 1934. His haemophilic elder brother, Alfonso, also died after a crash, in 1938.

With the death of Waldemar in 1945 the haemophilia gene which plagued the family life of royalty in three countries for almost 100 years and changed the course of European history, appears to have become extinct. There remains a possibility that the disorder could reappear in the descendants of one of Queen Victoria's daughters or of her only affected son, Prince Leopold.

Contemporary advances

Thirty years ago most people with haemophilia died from bleeding, often after operation, in childhood. Today they can expect to live a normal lifespan. Three major developments have been responsible for this remarkable change.

The first is the result of work of Professor Macfarlane and his colleagues in Oxford, England. By recognizing that a protein (factor VIII) is deficient or absent in haemophilia A, and that it could be replaced from normal fresh blood or plasma, they laid the foundation for modern haemophilia therapy. The second was the organization, mainly through the stimulus of world war, of the blood transfusion services, including those run by the Red Cross and Red Crescent, in many countries. Thirdly, has been the application of fractionation techniques on a commercial scale by American and European pharmaceutical companies. These techniques continue to play a vital role in the purification of clotting factors derived from both human and porcine plasma, and from recombinant DNA technology. They include essential steps designed to remove unwanted contaminants, including viruses and extraneous proteins, from the final product.

Over the years the blood products suitable for haemophilia therapy have become both more potent and more easily available, at least to people in developed countries. Fresh blood was followed by fresh frozen plasma and, in the early 1960s, by cryoprecipitate, discovered by Dr Judith Pool and her colleagues in Stanford, California. In parallel with these methods of treating blood from single donations, work was going on in many countries to find ways of obtaining concentrates from the pooled blood of many donors. Methods of fractionation were introduced by Cohn and Blomback and their associates in North America and Sweden, and quickly adopted by blood laboratories round the world. The introduction of freeze-dried concentrates by the pharmaceutical industry brought the benefits of really effective therapy to thousands of people with haemophilia, and allowed the development of large-scale programmes of home therapy and prophylaxis.

The World Federation of Hemophilia

1301 Greene Avenue
Suite 500
Montreal
Quebec
Canada H3Z 2B2

Tel: (010)–1–514–933 7944
Fax: (010)–1–514–933 8916

The World Federation of Hemophilia (WFH) was founded as a direct result of the work of one man, Frank Schnabel. Schnabel was born with severe haemophilia in Washington DC and lived in Montreal. Although his childhood was disrupted by repeated bleeds and little treatment he remained remarkably active, and with the encouragement of his mother eventually obtained a university degree. After helping to found the Canadian Hemophilia Society, Schnabel's persistence resulted in a congress of ten national haemophilia societies in Copenhagen in 1963, and the World Federation was born. Since that first meeting people with haemophilia and doctors have gathered regularly in cities throughout the world to discuss the changing patterns of haemophilia care. Frank died in 1987.

The WFH works closely with many other national and international organizations concerned with health and welfare, and is officially recognized by the World Health Organization. In the United Kingdom the address of the Haemophilia Society is 123 Westminster Bridge Road, London, SE1 7HR. In the United States the Hemophilia Foundation address is The National Hemophilia Foundation, 19 West 34th Street, Suite 1204, New York, NY 10001, USA.

Through its international secretariat, WFH is active in many fields of organization, research, and sponsorship. Perhaps the greatest of the tasks the Federation has set itself is to reach the many thousands of people with haemophilia in the poorer nations and to help them and their families live more normal lives. To this end the Federation's second President Charles Carman, an American businessman with severe haemophilia, launched a major strategic plan in 1990. The aim of this—the Decade Plan—is to try and bring help to all people with haemophilia in the world. This is in itself a daunting task and, without the possibilities of cheap recombinant material or genetic cure, an impossible one. But given the President's grit and determination and the drive of the task force of doctors and scientists he assembled from many countries, results are already being seen from the Plan. The initial impetus for this work was generously funded by the Armour Pharmaceutical Company, and given this sort of funding and goodwill from philanthropic and charitable donors in the future, there is now genuine hope for the world's haemophilic peoples. The present president is Brian O' Mahony from Ireland. Elected in 1994, he is determined to see the Decade Plan implemented.

The future

In the first edition of this book 20 years ago I wrote that the future promised early diagnosis of haemophilia in the womb, genetic manipulation of the fetus, the synthesis of clotting factors as a result of the mapping of their structures, and new ways in which the problem of inhibitors might be tackled. Ten years ago the predictions were for introduction of artificial blood products before 1994, chorionic villus diagnosis of haemophilia in the womb, oral factor VIII therapy, and better artificial joints. Four years ago, the promise of gene therapy seemed realistic.

But in the late 1980s the attention of most people with haemophilia and their doctors was still fixed on the safety of the products used for treatment. Elimination of viral contaminants and autologous blood transfusion were considered, as was the emergence of prophylaxis as a sensible alternative to on demand therapy. So fast have been developments in medicine that all these are now described in detail in other chapters. Other 'glittering prizes' are already within our grasp or on the way. Indeed, the coming years hold out more hope for people with haemophilia and their families than ever before.

Whilst in developed countries treatment itself has improved, and continues to improve, so has the infrastructure upon which medical diagnosis and monitoring depends. Automated machines are now sufficiently reliable and cheap enough to take over much of the tedious laboratory work involved in the diagnosis of bleeding and clotting disorders. High-definition ultrasonography ('ultrasound'), computed assisted tomography (the 'CAT scan'), and nuclear magnetic resonance imaging (the 'MRI scan') are revolutionizing the ease with which we can look into the body without surgery and diagnose bleeding and alterations in normal structure. Advances in the design of probes like the endoscope allow direct access for inspection and biopsy of body cavities without the need for operation. Laparoscopy is changing forever the whole field of surgery. 'Keyhole' incisions now take the place of operations that previously required major wounds and all that entailed in terms of time to heal and consequent use of valuable in-patient resources. These and other advances have allowed us to do away with much of the need for unpleasant and sometimes painful procedures.

Gene therapy

People with haemophilia and their families have long dreamt of the day when a cure might be found. This is particularly so for the many thousands of those with severe haemophilia A and B who live in developing countries and presently have no access to any treatment at all. To these people a permanent cure will be the only way to bring them relief. Both blood products and genetically engineered clotting factors are simply too expensive ever to provide a world-wide everyday answer to haemophilia.

Now the day when the dream turns into reality seems near. On September 14th 1990 in the United States the first human, a little girl born without normal body defences against infection, was partially cured by gene therapy. Since then other children with the same disorder, which is called adenosine deaminase or ADA deficiency, have been similarly treated in several countries. The doctors carrying out the techniques have moved forward very cautiously; the first gene therapy took eight years to plan and implement. This caution is necessary because of possible dangers inherent in trying to change someone's genetic make-up. Because no one knows what the long-term effects of gene therapy might be, or whether there may be unexpected harmful changes, work with humans can only follow animal experiments. This is now the stage at which gene therapy for the haemophilias has reached. In October 1993 doctors from Houston, Texas, and Chapel Hill, North Carolina announced that for the first time two dogs with haemophilia B (Christmas disease) had been given new genetic instructions, and that factor IX had subsequently been made in their bodies. The amounts of factor IX were very small but the experiment did show that the technique worked. What then is gene therapy?

The genes

All living things, both in the plant and in the animal kingdoms, need certain essentials for survival. Fundamental to these essentials which allow living things to exist, to defend themselves and to reproduce is a good set of instructions. These instructions are found in every living cell, and how the cell interprets them determines its form and its function.

Human beings have around 100 000 pieces of instruction; 50 000 are inherited from the father, and 50 000 from the mother. Each piece is called a GENE, and the total collection of genes in any one person is called the GENOME. It is your genome that determines that you are you!

The first complete map of the human gene was completed in Paris, and published in December 1993. The genome is organized on a series of structures called chromosomes. Humans have 23 pairs of chromosomes in each of their cells, with the exception of the eggs and sperm (the germ cells) in the ovaries or testicles. The genes on two of the chromosomes determine sex. These chromosomes are termed X and Y. Females are female because they have inherited two X chromosomes (XX). Males are male because they have inherited an X and a Y chromosome (XY). The genes which are concerned with the manufacture of clotting factors VIII and IX in the body are carried on the X chromosome. The Y chromosome has no clotting factor instructions on it at all.

That is why males have haemophilia and females do not. If a male has a faulty factor VIII or IX gene on his X chromosome he has no duplicate instruction to fall back on. If a female has a faulty gene on one of her X chromosomes there is a duplicate normal instruction on her other X chromosome (see Chapter 4).

The genome is the blueprint for life. Just as in a factory there are blueprints (or plans) for making a product, so in a cell there is the genome. In a factory the blueprint will be found in the office. In the cell it is found in the nucleus. From the nucleus the instructions are sent out into the cell to enable it to make things and to function.

Genetic instructions are inherited and work by using a clever yet straightforward code. Instructions are read from lengths of this coded message by structures within the cell, and end products assembled to the precise specifications given in the message. If something is wrong with the message a faulty product is made.

Genes are made from deoxyribonucleic acid (DNA). DNA exists in the nucleus in the form of a double helix. The precise sequence of the DNA acts as a template from which instructions are copied. These single-stranded instructions are called ribonucleic acid or RNA. The RNA carrying a set of instructions out from the nucleus into the cell factory is, not surprisingly, called messenger RNA (or mRNA).

Thus:

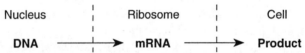

The RNA instructions are translated into products within structures on the factory floor called ribosomes.

Haemophilia is caused by a fault in the instructions given to the cell factory by one gene. If that gene could be corrected, and the right instructions for making normal clotting factor received by the cell, haemophilia would be cured.

To date, we know:

- which gene codes for factor VIII;
- which gene codes for factor IX;
- the precise structures of each of these genes;
- where they are located on the X chromosome;
- which cells in the body use the genetic instructions to make the clotting factor;
- most of the defects in those genes that cause haemophilia.

And we have already used our knowledge to introduce new genetic instructions into animal cells to make synthetic factor VIII and factor IX. Can we use the same technique to cure haemophilia? Let's first look at how we make recombinant clotting factor. A more detailed account of factor VIII manufacture will be found in Chapter 6.

Making factor VIII and IX

In order to make synthetic factor VIII concentrate a new sequence of instructions is produced artificially by joining pieces of DNA together. The resulting sequence is called recombinant DNA (or rDNA). In the case of factor VIII the product it produces is called recombinant VIII (or rVIII).

The newly created instructions are inserted into a living cell in the laboratory. Because the instructions for factor VIII are very large, a bacterial cell of the sort used for most genetic engineering is too small. Cells grown in culture from a mammal are therefore used, the particular cells in question originally being either from the ovary or the kidney of a Chinese hamster. Once in the cell nucleus the new DNA starts to instruct the cell factory via messenger RNA, and the factory starts to produce human factor VIII. The cells multiply and millions of them grown in the culture vats soon produce enough of the active protein for factor VIII to be extracted and concentrate made from it.

Another approach to producing a clotting factor using genetic engineering has been explored in animals. If the genetic recipe for a human protein is introduced into an animal the animal's cells manufacture the human product as well as its own proteins. It can then be extracted and used to control disease without worries about rejection as a non-human medicine. Using this technique the gene for human factor IX has been introduced into the milk-producing cells of sheep, which then produce small amounts of human factor IX in their milk, from which it can be harvested. Because the instructions are transferred from another species (in this case human to sheep) such animals are called transgenic.

Unfortunately, to date insufficient factor IX has been expressed in this way to make the process viable commercially. In order for that to happen ways must be found to enhance the output, or expression, of the factor.

One of the attractions of this sort of genetic farming is that the new instructions can be given to animals in such a way that the new characteristics are passed down to their offspring. This is called germ-line gene therapy because it involves manipulation of the germ cells of eggs or sperm. Whilst it would be possible in humans germ-line gene therapy is many years away. Before it can be used we must be absolutely sure that no harm will come by permanently altering even one gene. Problems may not show up for years and any fault present would be transmitted to future generations.

For these reasons gene therapy in humans is presently directed only to the individual. Because only a person's *body* is treated and not their sperm or eggs, the techniques used are called 'somatic' (*soma* = Greek for 'body').

Somatic gene therapy

For somatic gene therapy to work three things are needed. They are:

- the normal gene
- a way of getting it into a cell (a 'vector', or 'shuttle')
- a target cell, which will use the new gene to make and express the normal product

The normal gene

If an identical fault was present in everyone with haemophilia plans might be made to simply try and correct the mistake by deleting it, and substituting a correct piece of the instruction. This is not possible because most people with haemophilia A or with haemophilia B have widely differing genetic faults. Fig. 3.13 shows part of the factor IX gene with the sites of change from the normal sequence found in people with haemophilia B marked (see p. 52).

Change in only one item of the sequence is sufficient to result in haemophilia. Each single change of this sort is called a 'point mutation', and many different point mutations are now known in haemophilic families, especially in people with haemophilia B. In contrast rearrangements of the gene appear to account for about half the cases of haemophilia A (see p. 77)

It would be an impossible task to attempt to correct each abnormality on an individual basis. Far more likely to succeed for the greatest number of people would be to augment the faulty instructions by placing new, normal instructions alongside. The instructions the person had been born with would go on working (in this case producing useless clotting factor), but his haemophilia would be cured by the duplicate normal instructions.

In order to do this the normal gene must be introduced into his cells. In the case of factor VIII this gene and its resulting molecule are very big. This makes it difficult to squeeze the instructions down into a small enough space to put them into a possible vector. An alternative is to try and trim the gene sufficiently to allow it to be packaged without losing its ability to produce normal clotting.

This is presently done by removing large parts of the factor VIII gene called the B domain, apparently without harmful effect. A great deal of the human genome (most of it, in fact) seems to have no function, perhaps because it has become redundant with evolution. So there is good reason to believe that the factor VIII gene can be manipulated safely in order to make it fit its transport into the cell.

The vector

We already have highly efficient vehicles designed to go to the heart of our cells. They are the viruses and they cause numerous diseases from the common cold to AIDS. Viruses cause disease by highjacking cells using their own DNA. One of the most efficient members of the viral family is the retrovirus. Retroviruses are so-called because they work backwards (retro). Instead of the usual chain of events in which DNA → RNA, retroviruses persuade cells to accept RNA and convert it into DNA (RNA → DNA).

A retrovirus injects its own RNA into a cell. With the RNA is a chemical messenger or 'enzyme'. This enzyme, called 'reverse transcriptase', helps convert the single-stranded viral RNA into the double helix of DNA within the cell. This DNA is then inserted into the host's genome within the nucleus. From there the pirate DNA can send out instructions in the usual way, and hoodwink the cell to make copies of the original virus. In this way the infection spreads.

viral RNA → viral DNA → host DNA → mRNA → product
(new virus)

If we could insert our new factor VIII or IX gene into such a vector and infect a cell the new instructions would be delivered to the precise place we want them, into the genome which would then produce normal clotting factor.

First the retrovirus is made safe by engineering out its harmful qualities which would normally cause disease. Then the new gene is inserted, modified if necessary to make it fit. Then the vector is brought into contact with cells. This can be done outside the body after harvesting the right cells and then re-injecting the treated cells, or by direct injection into the body.

It was a retrovirus that was used as the delivery system in the dogs with haemophilia B described earlier. Before it could be used however, much of the dog's liver had to be removed. This ensured that there were plenty of dividing liver cells to take up the retrovirus.

Another viral vector used comes from the family of adenoviruses. In their natural form this group of DNA viruses cause sore throats and eyes, pneumonia, or diarrhoea. Again, suitably modified an adenovirus or adeno-associated virus can be used to transport a new gene into a cell, without causing illness. This form of vector has the advantage of not needing dividing cells in order to take root.

An exciting recent development has been the finding that the new gene might be incorporated within the chromosome numbered 19. If factor VIII or IX instructions can be inserted in this way there may be less chance of long-term harmful side-effects, as this particular chromosome is not known to contain any cancer-inducing genes (oncogenes).

Another way of getting the new genetic instructions into someone is to inject them directly either into specific cells or into the body. An example of the latter is the method now being developed to deal with cystic fibrosis, the new genes being administered to the lungs by aerosol breathed in intermittently by the patient.

The target cell

Much depends on the cells targeted because individual genes do not work on their own. They need regulating in order to ensure that they work properly. They need to be switched on when they are needed and switched off when they are not. Like most men they need housekeepers to look after them and keep them in control. All these functions are provided by other genes.

One reason why early attempts at gene therapy have not led to immediate returns to normality and therefore cure is because the output of the new genes needs to be enhanced, or promoted. To be successful the new gene must be put into a cell capable of carrying out its instructions. That cell must also be able to live a long life or to reproduce itself so that the effect continues. Without these attributes gene therapy cannot be permanent and will need to be repeated at intervals, although not as frequently as present therapy with clotting factors.

Several cells have already been targeted as suitable candidates for the introduction of gene therapy. As well as the lung for cystic fibrosis they include the bone marrow, which contains enormous quantities of early (stem) and dividing cells, and muscle. The two Chinese boys with haemophilia B are said to have received their gene therapy given in one form of muscle cell, the fibroblast. One of the advantages of using muscle is that the implant of treated cells can easily be removed in case anything goes wrong. Altered cells living deeper or more widely spread in the body will be harder to deactivate. One way devised to switch them off is to implant the new gene with another so called 'suicide gene'. This is triggered by giving a specific drug, causing the engineered cells to self-destruct.

As far as haemophilia A and B are concerned, in our present state of knowledge the best cells for genetic engineering are the liver cells called hepatocytes. It is the hepatocytes that normally produce factors VIII and IX, so they already contain all the regulatory mechanisms needed for the successful birth of normal clotting factors. Again in the dog experiments it was the hepatocytes that were targeted. A better target would be a liver stem cell which would then continue the line of new genetic instructions with a long-lasting effect.

Whilst these techniques look complicated they will not necessarily require sophisticated hospital arrangements in order to offer them to patients. As with haemophilia home therapy all that is needed is a specific, safe, and effective product and someone who can give it. Such a system is already working for the treatment of a range of diseases, including cancer, in several countries including the United States, Israel, and Switzerland. Cells removed in the blood from the patient at home are placed in a sterile container and taken to a 'cell hospital'. Here they are engineered and allowed to grow in culture. Specific attributes to enable them to combat the patient's disease are given to them and they are then returned by injection to the patient. Because all the cells are from his or her own body in the first place, the patient does not reject them as foreign. Gene therapy will shortly be one of the techniques added to a growing list of diseases they may be treated in this way.

Finally, in thinking of cure there may be no need to go to the extent of the forms of gene therapy described. We know that the transplant of a normal liver into someone with haemophilia cures the disorder. It does this because the transplanted cells already working in the organ contain the correct instructions for making factors VIII and IX. Provided the transplant is not rejected this provides a permanent cure. It is, however, a major undertaking and, of course, first a liver must be found and taken, for instance, from someone who has just died in an accident. But in the future it may be possible to provide enough factor VIII or IX to someone by implanting not a whole or even part of the liver. Instead a transplant just of active hepatocytes may be possible. This possibility is nearer with the discovery of the liver stem cell which should allow specific cultures of individual cells to be harvested, but it does need success in being able to overcome the problem of rejection by the recipient's defence systems. We need to know how to wrap such cell implants up to protect them, whilst at the same time ensuring that they continue to receive the oxygen and nutrients needed to ensure a suitably long life.

These developments in medicine are very exciting for everyone, not just those with haemophilia. They will revolutionize our lives and those of our families. One of the questions asked by people with rare disorders is whether the firms already committed to financing genetic engineering will be bothered with them. The answer must be yes, if only to enable research and development costs to be recovered.

As far as haemophilia is concerned gene therapy in the form of single gene or cell transplant is the only way forward. This was recognized by the Chinese who in one step hoped to advance from the days of little or no treatment to a cure, without having to try to develop blood transfusion services or recombinant clotting factor replacement therapy on the way. Perhaps their experiment will prove to be one small step for a Chinese boy with factor IX deficiency and one giant leap for mankind with haemophilia!

APPENDICES

Appendix A

A charter for people with haemophilia

If you have haemophilia or a related bleeding disorder, six basic needs should be answered. These are:

(1) accurate diagnosis;
(2) effective and safe treatment;
(3) twenty-four hour cover;
(4) regular follow-up;
(5) expert counselling; and
(6) a good standard of communication about your disorder.

Accurate diagnosis

Both the clotting factor affected and its level in your bloodstream should be known. You should be given a card with this information recorded on it, together with details of blood group, hepatitis B and C status, and presence of inhibitor, or allergy. Also recorded on the card should be the address and telephone number of your usual haemophilia centre. In the UK, such a card is issued by your haemophilia centre on behalf of the Departments of Health. The World Federation of Hemophilia card is similar. Always carry the card so that if an accident occurs those looking after you know the diagnosis. Alternatively wear an identity bracelet or pendant (Medic Alert or Talisman).

Effective and safe treatment

At present most people with severe haemophilia rely on blood products made from human plasma for their treatment. This position is changing. As they become more available after licensing the recombinant, genetically engineered products will take an increasing part of the market.

However, to be widely accepted the newer concentrates must come down in price. Present knowledge suggests that, in terms of safety and efficacy, there is little to choose between them and the high-purity plasma products. All raise the circulatory levels of the relevant clotting factor sufficiently fast and sufficiently high to prevent or to stop bleeding. Individual donor rejection and screening, together with antiviral treatment during manufacture, have made the newer plasma products safe. Most manufacturers are now implementing dual safety measures to ensure that all viruses are removed—for instance, a heating step is added to the monoclonal antibody process. These products are also often easier to prepare and give than their predecessors. They go into solution faster, and are made up with less volume of diluent.

There is no doubt, however, that some of the earlier products (sometimes called intermediate purity products) are equally safe in terms of virus transmission, and equally good in terms of haemophilia treatment. The arguments about the effects of proteins other than factor VIII or IX are by no means clear-cut. And some of the earlier products are the mainstay of treatment in some people with von Willebrand disease because they contain the von Willebrand factor.

What the prescription of a particular product comes down to in the end depends on whether it is licensed for use in a particular country and its cost, as well as patient preference. It is clearly silly to use the newest and most expensive product if it is of no additional benefit (or even marginal benefit) to the person using it.

Twenty-four hour cover

Much of this is provided by home therapy but there are occasions when a hospital visit is necessary. You and your family should know exactly who to contact at your nearest haemophilia centre, both by day and night. You should know how to arrange for an ambulance to be sent if other transport is not available. The staff of many hospitals will call the ambulance authority directly for you. Some people will need to ask their family doctors. Emergency numbers (of haemophilia centre, family doctor, nearest hospital, and ambulance) should be kept by the telephone at home.

The worst place for someone with haemophilia who is bleeding to attend for treatment is the Accident and Emergency (Casualty) Department. This is because most of the doctors working there will be comparatively junior and will probably not have seen someone with haemophilia before. They will not appreciate that treatment is indicted on your word and in the absence of any clinical evidence of bleeding. If you have to attend such a department regularly, and are not happy with the treatment you receive, do not be afraid to seek the advice of the Director of your nearest haemophilia centre.

If you go on holiday or on business away from your own area ensure that you know where to go for treatment. You are more likely to receive help quickly if you carry a letter of introduction with details of your treatment with you. In appropriate circumstances, you should also carry a holiday kit containing blood product, together with a signed letter to enable you to clear Customs with concentrate, needles, and syringes.

Regular follow-up

This is **vital** for anyone receiving blood products. Every severely affected person with haemophilia should have a physical check-up *at least once a year*.

You should be examined and your blood pressure and urine checked. Your blood should be tested for anaemia, white cell and platelet levels, liver function, including hepatitis A, B, and C status, clotting factory inhibitors, and immune status. When appropriate the latter tests should include human immunodeficiency virus (HIV) antibody detection. A physiotherapist should measure your joint ranges and muscle power, and X-rays of target joints should be taken. You and your family should see the results of the investigations and have time to ask questions about them.

Expert counselling

The constellation of questions that arise in any family with haemophilia can only be answered by a team of people expert in specific areas of medicine and surgery, social work, and nursing. If families feel that the answers are not available locally, they should seek advice at their nearest Comprehensive Care Haemophilia Centre. The address of this may be obtained from their family doctor or Centre, or from the Haemophilia Society, Hemophilia Foundation, or World Federation of Hemophilia (p. 346). Many countries have special arrangements for genetic counselling. Addresses may be had from family doctors, haemophilia centres, or consultant haematologists.

Good communication

Families should try to ensure that the day-to-day life of the person with haemophilia is not clouded by the disorder. Schools, workplaces, and neighbourhoods should have up-to-date knowledge of haemophilia so that no one is frightened by it, and everyone knows how to cope on the very rare occasions that a problem arises. Good communication works both ways. The person who keeps appointments and keeps healthy is more likely to be popular with his doctor than the man who does not bother.

In order to meet these needs the UK Departments of Health have recently issued an update on the criteria they expect to be met by the haemophilia centres. This updated information on the services provided by centres is reproduced with permission.

A Comprehensive Care Centre should provide:

1. a clinical service provided by experienced staff for the treatment of patients with haemostatic disorders and their families at short notice at any time of the day or night.
2. a laboratory service capable of carrying out all tests necessary for the definitive diagnosis of haemophilia and all common inherited haemorrhagic disorders, including the identification and assay of the relevant specific haemostatic factors. Further, capable of monitoring therapy and carrying out preliminary testing for inhibitors.
3. where appropriate and indicated, to conduct in collaboration with other haemophilia treatment centres, the further investigation of relatives of patients with haemophilia or other haemostatic disorders.
4. an advisory service to patients and close relatives on matters specific to haemophilia. Advice should also be given to general practitioners as appropriate.
5. maintenance of satisfactory quality control and assurance for all laboratory tests offered in relation to clinical services, both by establishing appropriate internal procedures and by participation at the appropriate level in the UK National External Quality Assessment Scheme in Blood Coagulation (NEQAS), or other relevant approved external quality assessment schemes.
6. maintenance of medical records; records must be maintained of all treatment administered and all adverse reactions reported. Special medical cards are to be issued and a register kept of all patients attending the centre.
7. counselling in privacy of patients and their relatives.
8. participation in appropriate clinical audit.
9. where appropriate, to provide advice on and organization of home therapy programmes either individually or in collaboration with other haemophilia treatment centres.
10. the provision of prophylactic treatment programmes for patients with haemophilia and other haemostatic disorders.
11. 24-hour advisory service to Haemophilia Centres and support to such Centres as appropriate.

---

12. a specialist consultant service for all surgery including orthopaedic and dental, for infectious diseases (such as HIV and hepatitis) and paediatric care, and for genetic, HIV, and social care and any other counselling services.
13. a reference laboratory service for Haemophilia Centres. The services should also include the diagnosis of atypical cases, genotypic analysis, the assay of inhibitors and other haemostatic factors, the diagnosis of hereditary platelet disorders, the supply of assay standards and reagents, and when requested, advice and recommendations concerning analytical procedures.
14. educational facilities for medical staff, nurses, medical laboratory scientific officers, counsellors and other personnel as required in order to promote optimal comprehensive care of patients.
15. co-ordination of meetings and undertaking research programmes, including the conduct of clinical trials and to establish and participate in suitable Regional and National programmes of clinical audit.

A Haemophilia Centre looking after fewer than 40 severely affected people should produce: (1) to (9) above.

Reproduced with permission from UK Departments of Health.

In reading this health circular, it must be remembered that haemophilia has to take its place with other disorders of the blood when it comes to provision of money and facilities. These disorders include leukaemia and other forms of cancer, and conditions that require bone marrow transplant. The dilemma of many haematologists faced with the problems of limited finance and time is obvious. Doctors welcome help from the charitable organizations set up to serve their patients. If your facilities are poor, work with your doctor and your colleagues to improve them!

Appendix B

Products containing aspirin

This section contains a list of compounds containing aspirin (ASA; acetylsalicylic acid) which should **never** be taken by anyone with a bleeding tendency. In naming aspirin preparations, the ingenuity of man is clearly infinite. In previous editions of this book I particularly liked 'Grandpa', but this name has now disappeared from the list.

Recently a serious illness called Reye syndrome has been linked to the use of aspirin in childhood. In the United Kingdom, this link has resulted in a recommendation that aspirin no longer be given to any child under the age of 12 years whether or not he or she has a bleeding disorder.

Aspirin (ASA; acetylsalicylic acid)-containing compounds which should not be taken

AAS
AB FE
AC & C
ACENTERINE
ACESAL
ACESOL
ACETAMINOPHEN and ASPIRIN TABLETS, USP.
ACETAMINOPHEN, ASPIRIN, and CAFFEINE CAPSULES USP.
ACETAMINOPHEN, ASPIRIN, and CAFFEINE TABLETS USP.
ACETARD
ACETISAL
ACETOPHEN
ACETYL
ACETYLIN
ACETYLO
ACETYLOSAL
ACETYSAL
ACIMETTEN
ACISAL
ACTISPIRINE
ACTRON
ACTRON COMPUESTO
ADDED STRENGTH PAIN RELIEVER
ADESINE AS
ADIRO
ADPIRIN

ADULT ANALGESIC PAIN
  RELIEVER
ADULT STRENGTH PAIN
  RELIEVER
ALASPINE
ALBYL
ALBYL-E
ALBYL-KOFFEIN
ALBYL-SELTERS
ALCACYL
ALEGRINA
ALGIMAX
ALGISPIR
ALGO
ALGO-NEVRITON
ALGOCRATINE
ALKA-SELTZER
ALKA-SELTZER EFFERVESCENT
ALKA SELTZER EXTRA
  STRENGTH
ALKA SELTZER FLAVORED
ALKA-SELTZER PLUS COLD
ALKA-SELTZER PLUS
  NIGHTTIME COLD
ALSOGIL
ALUPIR
ALUPRIN
ANACIN
ANACIN EXTRA STRENGTH
ANACIN MAXIMUM STRENGTH
ANACIN WITH CODEINE
ANADIN
ANADIN EXTRA
ANADIN MAXIMUM
ANADIN SOLUBLE
ANAFEBRYL
ANALGEN
ANALGESICO PYRE
ANALGIN FORTE
ANALVAL
ANCASAL
ANEXSIA
ANGETTES

ANODYNOS
ANTIDOL
ANTIGRIPPINE MIDY A LA
  VITAMINE C
ANTINEVRALGICO DR. KNAPP
ANTINEVRALGICO PENEGAL
ANTIREUMINA
ANTOIN
APAC
APC
APERNYL
APF
APO-ASA
APO-ASEN
APYRON
AQUAPRIN
ARTHRALGYL
ARTHRISIN
ARTHRITIS PAIN FORMULA
ARTRIA
ARTRIA S.R.
ASA
ASA
ASA-ASPIRIN
ASA and CODEINE COMPOUND,
  NO. 3
ASA ENSEALS
ASA with CODEINE
ASADRINE
ASADRINE C-200
ASAFERM
ASAGRAN
ASALITE
ASART
ASASANTIN
ASASANTINE
ASATARD
ASCRIPTIN
ASCRIPTIN EXTRA STRENGTH
ASCRIPTINE
ASDOL
ASKIT
ASPALGIN

ASPALOX

ASPASOL

ASPAV

ASPEC

ASPEGIC

ASPERBUF

ASPERCIN

ASPERGUM

ASPERMIN

ASPIDOL

ASPIGLICINA

ASPILISINA

ASPINFANTIL

ASPIRIN

ASPIRIN and CODEINE NO.
   2

ASPIRINA

ASPIRINA C

ASPIRINE

ASPIRINE C

ASPIRINETAS

ASPIRINETTA

ASPIRINETTA C

ASPIRISUCRE

ASPIRTAB

ASPIRVESS

ASPISOL

ASPRO

ASPRO C

ASPRODEINE

ASPROFARM

ASRIVO

ASS

ASS + C

ASS 500 DOLORMIN

ASS-KOMBI

ASSUR

ASTOPLEN

ASTRIN

ASTRIX

ATASPIN

AXOTAL

AZDONE

B-A-C

BABYPYRIN

BAMYCOR

BAMYL

BAMYL KOFFEIN

BAMYL S

BANIMAX

BAYASPIRINA

BAYER

BAYER 8-HOUR

BAYER CHILDREN'S

BAYER MAXIMUM

BAYROVAS

BC ARTHRITIS STRENGTH
   POWDER

BC POWDER

BC TABLET

BEBASPIN

BEBESAN

BEECHAMS POWDERS

BEECHAMS TABLETS

BENORAL

BERSICARAN N

BI-PRIN

BIONTOP

BISOLVON PILULES
   ANTIGRIPPALES

BISOLVON-GRIBLETTEN

BONAKIDDI

BOXAZIN

BOXAZIN PLUS C

BREOPIRIN

BREOPRIN

BRONCHO-TULISAN
   EUCALYPTOL

BUF-TABS

BUFACYL

BUFF-A

BUFFACYL

BUFFAPRIN

BUFFASAL

BUFFERED ASPIRIN TABLETS;
   USP

BUFFERIN
BUFFERIN, ARTHRITIS
  STRENGTH
BUFFERIN, ARTHRITIS
  STRENGTH TRI-BUFFERED
BUFFERIN EXTRA STRENGTH
BUFFERIN EXTRA STRENGTH
  TRI-BUFFERED
BUFFERIN TRI-BUFFERED
BUFFETS II
BUFFEX
BUFFINOL
BUTALBITAL AND ASPIRIN
  TABLETS; USP

C2
C2 BUFFERED
C2 BUFFERED with CODEINE
C2 with CODEINE
CAFASPIN
CAFERONAL
CAFIASPIRINA
CALMANTE MURRI
CALMANTE VITMDO P G
CALMANTE VITMDO PG
  EFER V
CALMANTE VITMDO RINVER
CALMANTE VTDO P G INFANT
CALMANTINA
CALMO YER
CALMO YER ANALGESICO
CAMA ARTHRITIS PAIN
  RELIEVER
CAMA INLAY-TABS
CAPRIN
CARDIPRIN
CARDOXIN PLUS ASA
CARIN
CARIN ALL'ARANCIA
CARISOPRODOL COMPOUND
CARTIA
CASPRIUM
CASPRIUM RETARD

CATALGINE
CATALGIX
CEBIOPIRINA
CEMIRIT
CEPHYL
CEREBRINO
CETASAL
CHEFARINE 4
CHEFARINE-N
CHEPHAPYRIN-N
CHILDREN'S CHEWABLE
  ASPIRIN
CHINASPIN
CHU-PAX
CHYMALGYL
CHYMOGRIP
CLARADIN
CLARAGINE
CLARIPRIN
CO-CODAPRIN TABLETS; BP.
COATED ASPIRIN
CODALGINA
CODALGINA RETARD
CODIPHEN
CODIS
CODOX
CODOXY
CODRAL BLUE LABEL
CODRAL FORTE
CODRAL JUNIOR
CODYL
COFETIL
COFFETYLIN
COJENE
COLDREX
COLFARIT
CONTRADOL
CONTRANEURAL N
CONTRE-DOULEURS
CONTRE-DOULEURS-C
CONTRHEUMA
CONTRHEUMA-RETARD
COPE

CORDEX
CORENZA C
CORICIDIN 'D'
CORICIDIN
CORICIDIN COLD TABLETS
CORICIDIN with CODEINE
CORYPHEN
CORYPHEN 325-CODEINE 30
CORYPHEN 650-CODIENE 30
COULDINA
CULLENS HEADACHE
   POWDERS

DAMASON-P
DAPRISAL
DARVON with ASA
DARVON COMPOUND
DARVON-N with ASA
DARVON-N COMPOUND
DASIKON
DASIN
DECRIN
DELGESIC
DESENFRIOL
DESENFRIOL C
DESENFRIOL D
DESENFRIOL INFANTIL
DESKOVAL N
DETOXALGINE
DIA-GESIC
DIA-GESIC IMPROVED
DIAFLEXOL
DIAFORIL
DISPERSIBLE ASPIRIN TABLETS
   BP.
DISPERSIBLE CO-CODAPRIN
   TABLETS BP
DISPRIL
DISPRIN
DISPRIN DIRECT
DISPRIN EXTRA
DOLEAN PH 8
DOLERON

DOLMEN
DOLOANA
DOLOFLEX
DOLOMEGA
DOLOMO TN
DOLOSARTO
DOLOVISANO
DOLOXENE COMPOUND
DOLPRN
DOLVIRAN
DOLVIRAN N
DOLVIS
DOMUPIRINA
DOSCAFIS
DOSTIL
DREIMAL
DRIN
DRINOPHEN
DRISTAN
DRISTAN EXTRA STRENGTH
DRISTAN FORMULA P
DULCIPIRINA
DURADYNE

EASPRIN
ECASIL
ECOPRIN
ECOTRIN
ECOTRIN EXTRA STRENGTH
ECOTRIN MAXIMUM
   STRENGTH
ECOTRIN REGULAR STRENGTH
ECOTRIN-10
ELSPRIN
EMAGRIN
EMCODEINE
EMCODEINE NO. 3
EMCODEINE NO. 4
EMPIRIN
EMPIRIN WITH CODEINE
EMPIRIN WITH CODEINE NO. 2
EMPIRIN WITH CODEINE
   NO. 3

EMPIRIN WITH CODEINE NO. 4
ENCAPRIN
ENDODAN
ENDOSPRIN
ENDYDOL
ENTERETAS
ENTERICIN
ENTEROSARIN
ENTEROSARINE
ENTROPHEN
ENTROPHEN 10
ENTROPHEN 15 MAXIMUM
  STRENGTH
EPROMATE
EPROMATE-M
EQUAGESIC
EQUAZINE M
ESKOTRIN
ETHICOD
EU-MED S
EUCALYPTINE ASPIRINE
  QUININE
EUCALYPTOSPIRINE
EXCEDRIN EXTRA-STRENGTH
EXTREN

FEMIDOL
FEMIGRAINE
FENSUM N
FINEURAL N
FIOGESIC
FIORINAL
FIORINAL CODEINA
FIORINAL with CODEINE
FIORINAL-C
FLECTADOL
FONAL N
FORTALIDON
FRIALGINA
FRIALGYL
FRIDOL
FRUMID
FYNNON CALCIUM ASPIRIN

GARASPIRINE
GELONIDA NA
GELPIRIN
GEMNISYN
GELPIRIN
GENSAN
GEPAN
GEPAN CN
GEYFRITZ
GLOBENTYL
GLOBOID
GLUCETYL
GODAMED
GOODY'S EXTRA STRENGTH
GRIPPAL

HA-TABLETTEN N
HAGEDABLETTEN
HALGON
HALPRIN
HEADSTART
HELVER SAL
HEMAGENE TAILLEUR
HEPTOGESIC
HERBIN-STODIN
HIPIRIN
HISTADYL and ASA
HYCO-PAP
HYPON
HYPRIN

IDOTYL
ILVICO
INYESPRIN
ISTOPIRINE
IVEPIRINE

JURIDIN
JUVEPIRINE
JUVEPIRINE A LA
  PROMETHAZINE

KALCATYL

KILIOS
KLAR
KREUZ-TABLETTEN
KYNOSINA
KYOWAMIN D

LABOPRIN
LAFENA
LASAFORT
LEVIUS
LICYL
LONGASA
LORTAB ASA

MAGNAPRIN
MAGNECYL
MAGNECYL-KOFFEIN
MAGNYL
MALEX
MALINERT
MANDROS
MAPRIN
MAXIPRIN-15
MEASURIN
MEDISYL
MEGA-DOLOR
MEJORAL
MELABON
MELABON PLUS C
MEPRO ANALGESIC
MEPROGESE
MEPROGESIC
MEPROGESIC Q
MICRAININ
MICRISTIN
MIDOL
MIGRANE-DOLVIRAN
MIGRAVESS
MINIMAX
MIOPHEN
MONACANT
MONOBELTIN
MORPHALGIN

NEDOLON A
NEO NISIDINA
NEO-CIBALGIN
NEO-CIBALGINA
NEO-CORICIDIN
NEO-GEPAN
NEO-NEVRAL
NEOCIBALENA
NEODONE
NEOGESIC
NEOLIN
NEOPIRINE
NEOPIRINE-25
NEOPYRINE
NEURALGIN N
NEUTRACETYL
NEVRAL VITAMINE B(1)-B(6)
NIBOL
NORGESIC
NORGESIC FORTE
NORWICH ASPIRIN
NORWICH EXTRA STRENGTH
NOVA-PHASE
NOVA-PHASE 5
NOVACETOL
NOVASEN
NOVID
NOVOPROPOXYN COMPOUND
NOVOSPRIN
NU-SEALS ASPIRIN
NURSE SYKES POWDERS

OCCIGRIP
OKAL
OKAL INFANTIL
ORPHENGESIC
ORRAVINA
ORTHOXICOL
OXYCODAN
OXYCODONE and ASPIRIN
    TABLETS, USP

PANODYNES ANALGESIC

PARAFLEX COMP.
PAYNOCIL
PERCODAN
PERCODAN-DEMI
PERSISTIN
PHARMACIN
PHENAPHEN NO. 2
PHENAPHEN NO. 3
PHENAPHEN NO. 4
PHENAPHEN with CODEINE
PHENAPHEN with CODEINE
    NO. 2
PHENAPHEN with CODEINE
    NO. 3
PHENAPHEN with CODEINE
    NO. 4
PHENETRON COMPOUND
PHENSIC
PIRECILINA
PLATET
PMS-ASA TABLETS
POWERIN
PRAECIMED N
PRAECINEURAL
PREMASPIN
PRENOXAN AU
    PHENOBARBITAL
PRESALIN
PRIMASPAN
PRODOL
PROPOXYPHENE COMPOUND
PROPYRE T
PROTECTIN-OPT
PROTEOSULFAN
PROVOPRIN
PYRACYL
PYRONOVAL

Q-GESIC
QUADRONAL AS
QUINTON

RECTOSAL YL

REFAGAN
REFAGAN N
RESPRIN
RESPRITIN
REUMYL
RHEUMYL
RHINOCAPS
RHODINE
RHONAL
RHUSAL
RIANE
RING N
RIO-JOSIPYRIN
RIPHEN
ROBAXISAL
ROBAXISAL FORTE
ROBAXISAL-C
ROBAXISAL-C 1/2
ROBAXISAL-C 1/4
ROBAXISAL-C 1/8
RODINA
ROMIGAL N
RONPIRIN APCQ
ROXIPRIN
RUMASAL
RUMICINE

SAL
SAL-ADULT
SAL-INFANT
SALABUFF
SALATIN CAPSULES
SALETO
SALICILINA
SALITISON
SALITISON C
SALOCOL
SANOCAPT
SANSPOMIN
SANTASAL
SARGEPIRINE
SASPRYL
SEDALGIN

SEDASPIR
SEDERGINE
SERUBON
SERVISPRIN
SIGURAN N RETARD
SILENTAN
SINASPRIL
SINE-OFF
SK-65 COMPOUND
SK-OXYCODONE with ASPIRIN
SODOL COMPOUND
SOLCODE
SOLMIN
SOLPRIN
SOLPYRON
SOLUSAL
SOLUSPRIN
SOLUVER
SOLVIN
SOMA COMPOUND
SOMA COMPOUND WITH
  CODEINE
SOPARINE
SPALT
SPALTINA
SPIRAMON
SPREN
SRA
ST JOSEPH COLD TABLETS
  FOR CHILDREN
SUPAC
SUPASA
SUPERASPIDIN
SUPPOTHERA A
SUPRIN
SYNALGOS-DC

TALWIN COMPOUND
TASPRIN-SOL
TECNAL
TECNAL C
TECNAL C 1/2
TECNAL C 1/4

TEMAGIN
TEMAGIN ASS
TEMPIL N
TENOL-PLUS
TENSTON
THOMAPYRIN
THOMAPYRIN C
THOMAPYRINE
TOGAL
TOGAL ASS
TOPTABS
TRANCOGESIC
TRANQUIGESIC
TREO
TREO COMP
TREUPEL
TREUPHALIN
TRI-PAIN CAPLETS
TRIAPHEN
TRIAPHEN-10
TRIGESIC
TRINERAL
TRIPLEX
TRIPLICE
TROMBYL

ULTRAVIRO C
UNIDOR
UPSALGIN C
URSINUS INLAY

VALESIN
VANQUISH
VEGANIN
VEGANIN 3
VEGANINE
VENOPRIN
VERALGIT
VERDAL
VERIN
VIA MAL
VIVIN C

WESPRIN
WESPRIN BUFFERED
WINSPRIN

YAFIN

ZORPRIN

Index

appliances 186–92, 259–61
arthritis 105–8, 110, 196
  anti-inflammatory drugs 273–4
  operations 196–202
  pain 270–6
  radionuclide synovectomy 201–2
arthrodesis 200–1
arthropathy, see arthritis
arthroplasty 198–201
arthroscopy 201
artificial insemination 310–11
artificial joints 199, 201
ASA, see aspirin
ascorbic acid deficiency 44
aspiration of joints 143, 241, 258
aspirin 16, 21, 35, 95, 272, 365–74
aura, haemophilic 142
Australia antigen/antibody 220–1
autoimmune disorders 206
Autoplex 211
autosomes 64
autosomal dominant disorders 78–80
autosomal recessive disorders 80–3
AZT, see zidovudine

babies 12–18
  sex of 67–8
baby bouncers 16
baby walkers 16
baby-sitters 15, 25
baby teeth 277
back muscles, bleeding into 99–102
ball and socket joints 106–7
battered babies 14
B cells 235
BCG vaccination 16, 269
Bethesda inhibitor units 207
bicycles 22, 288–91
birth 13, 310
bleeding: acute episodes 176
  clotting and 39–58
  prolonged 91
  spontaneous 97
  time 43, 83
  see also bleeds
bleeds 91–116
  bruises 6, 14, 17, 92
  closed 97–115

and early treatment 142
open 92–7, 141
recognition 6, 91
sites
  anus 17–18, 95
  back 99–102
  bowel 95, 148–9
  calf 104, 144
  chest 92–3
  epistaxes 92, 146
  eye 92, 145
  face 92, 145
  forearm muscles 99, 144
  fractures 108–10
  frenum 92
  gullet (oesophagus) 93–5
  haemarthroses 105–10, 143–4
  hamstrings 103–4, 144–5
  head 7, 17, 19, 36, 111–15, 146
  iliacus, see psoas
  joint 105–10, 143–4
  kidney 95–7, 147
  mouth 92, 145
  muscles 98–105, 144
  neck 145
  nervous system 111–15, 146
  nose (epistaxes) 92, 146
  piles 95
  psoas 100–2, 145
  quadriceps 103–4, 144–5
  scalp 92
  skin 6, 14, 17, 92
  skull, see head
  stomach 93–5, 148–9
  thigh 99–104, 144–5
  tongue 92, 145
  umbilical cord 116
  water (urine) 95–7, 147
  womb 147
treatment of 141–72, 176
blood
  background information 40–2, 234–6
  cross-matching 41–2, 197
  donation 119–22, 227–8, 262–3
  platelets 42
  red cells 41–2
  safety 119–22, 127–9, 227–30, 262–3
  separation of 122–7
  shelf life 121

LIBRARY
TOR & SW COLLEGE
OF HEALTH / N FRIARY
GREENBANK TERRACE
PLYMOUTH PL4 8QQ